*Measure what is measurable,
and make measurable what is not so.*

attributed to Galileo Galilei
1564-1642

Physics

Volume One
Classical Foundations

Rajesh R Parwani

Simplicity Research Institute, Singapore
www.simplicitysg.net

Physics
Volume One: Classical Foundations
Published by the Simplicity Research Institute, Singapore
www.simplicitysg.net
email: enquiry@simplicitysg.net

A CIP record for this book is available from the
National Library Board, Singapore.

ISBN: 978 − 981 − 11 − 3362 − 6 (pbook)
ISBN: 978 − 981 − 11 − 3363 − 3 (ebook)

Contents

I do not know what I may appear to the world,
but to myself I seem to have been only like a boy,
playing on the sea-shore,
and diverting myself in now and then finding a smoother pebble,
or a prettier shell than ordinary,
whilst the great ocean of truth lay all undiscovered before me.

Isaac Newton
1643-1727

Preface

This book is a concise survey of the foundations of classical physics. It focuses on conceptual issues, and the various limitations that were later overcome with the development of quantum theory and Einstein's relativity.

The presentation is aimed at enthusiasts in schools and beyond who have had some prior exposure to physics. However, the uninitiated might also find some parts of this book to be informative.

Notes, exercises, and references have been included for those who are more inquisitive. Additional resources are on the book's webpage

www.simplicitysg.net/books/physics.

Some sections of this book have been reproduced from Ref.[1] and Ref.[2]; the first reference contains numerous quantitative problems on the physics concepts discussed here.

Feedback from users of this book is most welcome. Please email **enquiry@simplicitysg.net**.

0.1 Conventions

Keywords are highlighted in *italics* while footnotes[1] provide clarification. Single 'quotes' focus on particular words.

The word *system* is used to refer to the limited part of the universe that we wish to study. Hence the universe is conveniently divided into two parts, the system and an exterior *environment*.

Most systems are *open*, allowing for interaction (an exchange of energy, matter or information) between the system and the environment; some systems may be approximated as *closed*.

Notes and References are numbered like this [2] and placed at the end of the book.

---

[1]Like this.

0.2 Acknowledgements

I am extremely grateful to Luke Gompertz, Gwendolyn Regina Tan, Lim Mei Ying, Thong May Han, Chan Zi Keane, Moses Khoo, Michael Cassidy, Anh-Minh Do and Elizabeth Cassidy, for providing helpful feedback on the draft manuscript.

I am also grateful to the hundreds of students who participated in the 'How Technologies Work' and 'Engineering Physics' modules that I taught between 2001 and 2012, and to the dozens of research students who joined me in physics adventures. They all helped to evolve the material that has been partly condensed into this book.

Rajesh R Parwani
May 2017
Singapore

The scientist does not study nature because it is useful;
he studies it because he takes pleasure in it,
and he takes pleasure in it because it is beautiful ...
I do not speak of that beauty which strikes the senses ...
What I mean is that profounder beauty
which comes from the harmonious order of its parts,
and which a pure intelligence can grasp.

Henri Poincaré
1854-1912

1

The Key Ideas Summarised

1. The *scientific method* is the basis for *physics*, which aims to understand the nature of space, time, matter, and their interactions.

2. Matter is composed of *atoms*, which contain electrically charged constituents.

3. Precise *equations* of classical physics describe the time evolution of a system, relating *cause* to *effect*.

4. The laws of physics take the same form in all *inertial* frames of reference.

5. *Symmetries* of the equations of physics are related to the existence of *conservation laws*.

6. The *entropy* of a closed system increases or remains the same. This defines the *thermodynamic arrow of time*.

7. *Spontaneous symmetry breaking* creates macroscopic *order* without any specific rules to that effect at the microscopic level.

8. Electric *charges* create electromagnetic *fields* in their vicinity, which then exert forces on other charges.

9. *Waves* allow for the transmission of energy, and hence information, without a net transfer of matter.

10. Many physical laws and classical concepts are known to be *emergent* from more fundamental entities.

Your Notes

Archimedes of Syracuse
287-212 BCE

A pre-eminent scientist, he was one of the first to use
mathematics for the study of physical phenomena.

2

What is Physics?

Physics is a scientific discipline that enquires into the nature of space, time, matter, and their mutual interactions.

But what is a *science*? The early stages of a science involve the observation of phenomena, the gathering of data regarding those phenomena, and an attempt to organise the information in a form that exhibits patterns. Ideally, as the science progresses, one hopes to have a deeper understanding of the data beyond what is observed; this is facilitated by the building of *models*.

In this chapter, we briefly discuss the model building philosophy and the importance of the scientific method.

2.1   The Scientific Method

A scientist seeks to find the most economical and accurate description of phenomena using the *scientific method*: testing predictions from theoretical models against data from experiment or observations of actual events.

The word *model* usually refers to a tentative framework to explain observed phenomena. Once a model has been tested and developed to sufficient generality, it is often called a *theory*[1].

The scientific method is not a linear process but involves many interlinked and iterative steps in the search for understanding. Typically, one first has a phenomenon in need of an explanation. The phenomenon might exhibit some regularities that may be summarised by empirical relations. Experiments (or observations) might be conducted to check the robustness of the phenomenon under controlled

---

[1]Unfortunately, the terminology is not standard.

conditions. A *hypothesis*, based on some model, may then be formulated: this is just a guess as to the cause of the phenomenon. Predictions can be made based on the hypothesis, and further experiments conducted to test them.

In other words, the scientific method involves both *inductive reasoning*, whereby a hypothesis or model is formed based on the available data, and *deductive reasoning* to reach a logical conclusion from the hypothesis.

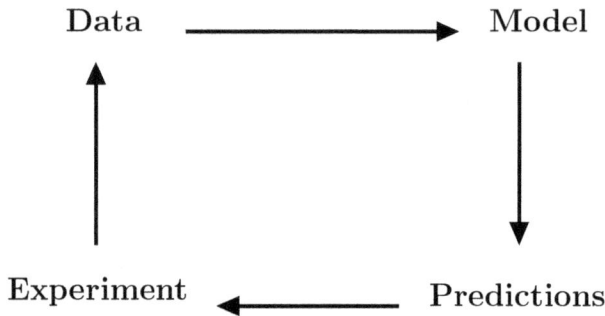

Figure 2.1: Ingredients in the scientific method.

Sometimes, the initial attempted theoretical explanation of a phenomenon might turn out to be incorrect (for example, Kepler's attempt to explain the origin of his laws), but if the data is robust, then that particular failure does not invalidate the entire scientific enterprise.

Other times, more than one model might be able to explain the available data without a discernible difference. In that case, one typically invokes *Occam's Razor* to support the simplest explanation over the more convoluted (A related idea in model building is the *KISS principle*: Keep it Simple, Scholar!). However, all such support is tentative, until more exploration strengthens our case or causes us to revise our views.

Some distinguishing features of the scientific method are:
(1) *Falsifiability* of the hypothesis — that is, one should be able to test the hypothesis.
(2) *Reproducibility* of results — the same experiment conducted by independent examiners under the same conditions should give com-

patible results[2].

2.2 Examples of the Scientific Method

1. The ancient Greek philosophers tried to use their logical and argumentative abilities to summarise the workings of the world in terms of a few concepts. Sometimes this natural philosophy led to surprisingly accurate deductions, but more often than not the speculative reasoning led the natural philosophers to misleading or unhelpful conclusions such as: "a dropped stone fell straight to the ground because that was its natural state, whereas a feather fell much slower and in an erratic manner because that too was its nature".

   It would take many years before Galileo would argue that both the stone and feather would fall at the same rate, and straight down, once the wind and air-resistance were neglected. This was a crucial development because it illustrated that profound facts can be deduced if we focus on simple things. *Controlled experiments* are done whereby only those factors one is interested in studying are allowed to vary while the rest are kept fixed.

   Such controlled experiments are idealisations of natural phenomena and are meant to uncover the underlying rules of how the world works. The use of controlled experiments, and the scientific method of comparing hypotheses and theoretical predictions with experiments, has been the key to the success of the natural sciences.

   By uncovering the few fundamental laws behind diverse phenomena, the world appeared to become more understandable and simple. With regard to the example above, Newton later showed that not only did the stone and feather fall towards the Earth at the same rate, so did the Moon: the law of gravitation is universal. This *universality* of fundamental laws is what makes us believe in the ultimate *simplicity* of the world[3].

---

[2]In quantum mechanics the outcome need not be identical each time but must nevertheless follow the predicted statistical law.

[3]See also Sect.(2.5).

2. When Mendeleev listed the known chemical elements in order of increasing atomic mass, he noticed that elements with similar chemical properties occurred at regular intervals. This led him to formulate the Periodic Table, which had several gaps filled by later discoveries.

   The understanding of the observed patterns would come later once the nature of the atom, in particular the arrangement of electrons in orbitals and their role in chemical properties, was established.

   This example illustrates the importance of recognising patterns in data during the early stages of a scientific investigation.

3. Since the earliest times, we have perceived temporal regularities in nature: day to night and day again, the recurring seasons, the motion of the Moon and other planets. Ptolemy held that the Earth was the immobile centre of the cosmos with the Sun and other planets revolving around it. Later Copernicus proposed a simpler scheme: the Sun at the centre with the planets revolving in circular orbits around it. The two proposals made different predictions that could be tested once accurate measuring instruments were invented to quantify the observations.

   Kepler improved on Copernicus' model, replacing the circular orbits by elliptical ones, and deducing a relation between the periods and size of the orbits. However, Kepler's laws were purely empirical, without any underlying explanation. That would come later with the work of Newton and his law of universal gravitation: Kepler's laws are now understood to be consequences of Newton's more fundamental laws which also explain terrestrial phenomena.

   However, even the great Newton had to bow to improvement when Einstein's General Theory of Relativity was verified by experiments and found to give a more accurate account of nature.

   This example illustrates how our understanding of the world typically improves and is refined as we get more information about it through technological progress — which in turn depends on our improved scientific understanding.

4. Around 1910, it was found that the neutron apparently decays according to the formula

$$\text{neutron} \xrightarrow{?} \text{proton} + \text{electron}. \tag{2.1}$$

Unfortunately, the observed process seemed to violate energy conservation, leading some physicists of that time to suggest that energy conservation was not an exact principle!

However, in 1930 another possibility was suggested by Pauli. He postulated the existence of a new particle that was named the neutrino: a third particle on the right-hand-side of the above equation would account for the missing energy. To fit the experimental data, this particle had to be massless, and hence move at the speed of light.

The neutrino was indeed found by Reines and Cowan many years later in 1956, with the exact properties required to conserve energy.

This example again illustrates the scientific method summarised in Fig.(2.1): observing a phenomenon, forming a hypothesis (energy must be conserved), making deductions (an undetected particle carrying away some energy), and experimentation. It also highlights that there can be multiple hypotheses, and it usually takes a long time before any of the hypotheses can be confirmed.

2.3 Measurement

Measurement is an essential part of doing physics. The metric units for length, mass and time are the metre (m), kilogramme (kg) and second (s). The prefix *kilo* (k) stands for 10^3; one kilogramme is 1000 grammes, and one kilometre is 1000 metres. Other common prefixes for large quantities are *Mega* (M) for 10^6 and *Giga* (G) for 10^9.

For small quantities, the common prefixes are *milli* (m) for 10^{-3}, *micro* (μ) for 10^{-6}, and *nano* (n) for 10^{-9}.

Some derived units of measurement have unique abbreviations. For example, since from Newton's Second Law force equals mass times acceleration, the unit for force is kg m s^{-2}, which is called the *Newton*

(N). Both force and acceleration are *vector* quantities, so one needs to specify their direction in addition to their magnitude[4].

Avogadro's constant, N_A, is 6.022×10^{23}; it is the number of atoms in 12 grammes of Carbon-12. The *mole* refers to an amount of substance which has N_A constituents. So, 1 mole of Carbon-12 has a mass of 12 g, while one mole of water has N_A molecules with a total mass of 18 g.

The *scientific notation* is very useful for expressing numbers compactly. Any non-zero number may be written in the form $\pm A \times 10^n$ where $1 \leq A < 10$ and n is an integer (positive or negative).

The number of digits in the A of the scientific notation expresses the *significant digits* for that quantity. For example, the distance from the Earth to the Moon is about 384,000 km. This may be written as 3.84×10^5 km or 3.84×10^8 m, expressing the distance correct to 3 significant digits (if we are confident of the value). If we were less certain of the last digit, we could express the distance as 3.8×10^8 m, signifying a precision of 2 significant digits.

The power, n, of 10 in the scientific notation is useful for comparing the relative sizes of numbers. If two numbers of the same sign have the same power in standard notation, then we say they are of the same *order of magnitude*. One order of magnitude is one power of ten; a number with $n = 5$ is two orders of magnitude larger than a number with $n = 3$.

For example, the distance of the Sun from Earth is about 1.5×10^8 km, which is about three orders of magnitude larger than the distance of the Moon from Earth.

2.4   Creating Mathematical Models

Mathematics is a precise language with an inbuilt logic. So it is ideally suited for the scientific study of patterns and trends, especially in physics which involves quantifiable[5] observables.

---

[4]Technically, a vector is a quantity whose components transform in the same way as those of the position vector **r** under rotations of the coordinate system. In contrast, quantities such as mass or temperature, which are specified solely by a magnitude, are called *scalar* quantities. Vectors are indicated by boldface, and their magnitude by the same symbol in normal font. A one-dimensional vector can be represented by a real number whose sign denotes direction.

[5]Some of the mathematics used in this book is summarised in Sect.(2.8).

Mathematical models usually start with several approximations and simplifications to capture the essential aspects of a phenomenon in a manageable form, and to allow for systematic improvements to the model. Of course, deciding what is vital in a model is an art!

For example, many early explorers modelled the Earth as a flat surface. That model was later replaced by that of a sphere, which was further improved to a squashed sphere (with the diameter in the North-South direction smaller than in the East-West direction).

Although we now know that the Earth is far from spherical, the simpler spherical model is still a useful approximation in some contexts.

In some situations, either for convenience or when a comprehensive theory is not yet available, two or more models might be used to describe different aspects of the same system. For example, it is sometimes convenient to think of light as a stream of particles, and at other times as a continuous wave.

A common error is to forget the limitations of a model and to use it where it is not valid[6]; a related error is to confuse the model with the actual phenomenon that is being modelled.

2.5   Information Compression

Humans seem to be good at discovering patterns in nature. It could be argued that regularities in nature had to exist for long periods of time so as to provide stable conditions for complex life forms to evolve. The evolving organisms would then have had an adaptive advantage if they could use the information they gathered from their surroundings to make predictions of the future.

While we have created different modes of communication to share our knowledge, the language of mathematics has proven its distinct advantage in physics. In this regard, the various mathematical structures and tools that have been invented are constrained only by their self-consistency, and their continued popularity depends on their utility.

The mathematical laws of physics may be viewed as a way of compressing information about the regularities of nature, just as

---

[6]For example, flat Earth, spherical Earth, and ellipsoidal Earth models each have their limited use.

a computer algorithm summarises the working of a computer pro-
gramme. Compressing more information, and in a more efficient
manner, achieves the goals of *universality* and *simplicity*.

The *complexity* and *diversity* that is seen in the world is then a
result of those laws operating on systems of different sizes, or with
different initial conditions [2].

2.6   Making Estimates

It is often useful to make estimates, or have some qualitative expecta-
tions, before constructing a detailed mathematical model. We discuss
some of the estimation techniques below and in the exercises.

2.6.1   Order of Magnitude

Making order of magnitude estimates is useful when we do not yet
have detailed information about a problem. For example, how many
hairs are on your head? You could attempt to count them all, with
some effort and much expenditure of time (during which some hairs
might fall off). But before starting, could you make an estimate? Is
the number likely to be in the region of 10^3, or 10^6, or something
else?

For an estimate, you might count the number of hairs in a small
region, determine the area of your head which has hair, and then
combine those pieces of information to get your answer. From this
result, you would also be able to estimate how long it would take to
count all the hairs individually.

2.6.2   Size and Similarity

Here we illustrate a simple but powerful estimation technique which
uses the idea of *scaling* or *similarity*.

Consider a solid cube of edge length L. Now imagine doubling
each edge to $2L$ while keeping the material density the same. Clearly,
the scaling by a factor of 2 increases the surface area by a factor of 4
and the volume by a factor of 8.

In fact, as the edge length is doubled, any cross-section of the
cube would increase its surface area by a factor of 4, and any three-

dimensional region of the cube would increase its volume by a factor of 8.

More generally, by dividing any region into a sum of infinitesimal regions (triangles or cubes), you can convince yourself that when the linear dimensions of any solid are scaled by a factor of k, areas scale by k^2 and volumes by k^3.

If the density remains the same, it would mean that a cube scaled by a factor of k would have k^3 times the weight of the original cube, and the pressure on its base would be $k^3/k^2 = k$ times the original pressure. Since real materials have finite strength, this means that scaled up versions of small objects might not be able to maintain their shape. For example, an ant that is scaled up to the size of an elephant would probably not be able to support its weight — that is why elephant legs are much thicker than those of a scaled up version of a smaller animal.

Here is another example: an ant can easily survive a fall from a great height but not a human. This is because air-drag, which slows down an object, is proportional to the surface area which scales as L^2, but the gravitational force is proportional to the mass of the object and scales as L^3 (for constant density). The larger creature will thus reach the ground with a bigger, and potentially lethal, velocity [1].

2.6.3 Dimensional Analysis

Dimensional analysis is a useful tool for obtaining relationships among the possible variables that might contribute to a measured quantity.

An equation like Newton's Second Law, $\mathbf{F} = m\mathbf{a}$, involves quantities which have different *dimensions*: m is a mass, so it is said to have the dimension of Mass; $\mathbf{a}$ is an acceleration, which can be measured in the units m/s^2 and is said to have the dimension of Length/(Time)^2.

Note that acceleration may be measured in other physical units, such as km/h^2 but its dimension will still be Length/(Time)^2.

Let us denote the dimensions of Mass, Length and Time by the letters M, L and T, and use the notation $[X]$ to denote the dimension of a quantity X. So, for example, [force] $= MLT^{-2}$.

Some quantities are dimensionless, such as angles measured in radians (since that is the ratio of two lengths), and pure numbers such as $-1/2$ and π.

Consistency demands that the dimensions of both sides of an equation match, and this is a quick way of detecting errors. For example, suppose you came across an equation which claimed to give the height h, reached by a projectile fired upwards with an initial speed v, as $h = g/v^2$. Could this equation be correct?

Let us check dimensions. The left hand side has dimensions $[h] = L$ while the right hand side has dimensions $[g/v^2] = [g][v]^{-2} = LT^{-2}(LT^{-1})^{-2} = L^{-1}$. The dimensions do not match, and so the equation cannot be correct.

Notice that the alternative equation $h = v^2/g$ is dimensionally consistent, but this does not imply that it is correct! In fact, the correct equation is $h = v^2/(2g)$ (when air resistance is negligible).

So we see that dimensional analysis may be used to detect errors but it does not, by itself, guarantee the correctness of an equation.

Dimensional analysis may be used constructively in the model building process. As an illustration, suppose that we did not know how to derive the formula for the period of a simple pendulum. Could we still deduce something about how the period depends on the possible variables?

First, we need to make a guess about the likely variables that might influence the period P: the mass of the bob m, the length of the string l, the acceleration due to gravity g, and the starting angle from the vertical A. Let us then suppose a relation of the form

$$P = Cm^w l^x g^y A^z \tag{2.2}$$

where C, w, x, y, z are constants to be constrained by dimensional analysis[7]. Note that $[P] = T$, $[C] = [A] = 1$ (dimensionless). So the right-hand-side of our guessed relation Eq.(2.2) has the dimensions $M^w L^x (LT^{-2})^y = M^w L^{x+y} T^{-2y}$. Since this must equal T, we deduce that $w = 0$, $x = -y = 1/2$.

So we conclude that the formula for the period should be of the form $P \propto \sqrt{l/g}$. Even though the proportionality constant might depend on the dimensionless amplitude A in some complicated way (in fact it does), we were still able to deduce a simple dependence of the period on the length of the string — without solving a differential equation, see Sect.(4.5).

---

[7]C might also depend on dimensionless combinations of some of the variables.

Sometimes one must include constants in the analysis, such as the gravitational constant G, that have a dimension (see exercises below).

2.7 Summary

At any moment in time, we are likely to have only partial information about a phenomenon. We use current principles and models in an attempt to explain that phenomenon and to push the boundaries of our knowledge. If new, verified, observations do not support the existing framework, then the latter must be modified to provide an even better and more encompassing description of nature.

Creativity plays a significant role in the practice of science, in the formation of various subjective hypotheses and models. But the scientific method is a crucial constraint: the theoretical constructs must be validated by empirical data. In this way, science achieves its essential objectivity.

Publishing unproven hypotheses, which may later turn out to be incorrect, is part of the scientific process. But, being human, scientists also have all the usual failings which can lead to mistakes, especially when there is a rush to publicise and claim priority. As such, even papers that are published in reputable peer-reviewed journals may turn out to be wrong, either due to oversight or other reasons (explicit fraud does occasionally occur).

However, on the whole, sooner or later, independent critical analyses by other scientists detect and correct such errors: while individuals scientists might fail, the collective enterprise is able to progress.

2.8 Appendix: Some Mathematics

Some of the mathematics notation used in this book is summarised here; more is in Ref.[3].

1. Given a sequence of numbers a_i with the index i taking values from 1 to N, denote their sum by

$$\sum_{i=1}^{N} a_i \equiv a_1 + a_2 + \ldots + a_N \ . \tag{2.3}$$

This will sometimes be condensed to $\sum_i a_i$.

2. The trigonometric function $\sin(x)$ is *periodic*, $\sin(x) = \sin(x + 2\pi)$, and odd, $\sin(-x) = \sin(x)$. It ranges between -1 and 1 with $\sin(0) = \sin(\pi) = 0$ and $\sin(\pi/2) = -\sin(3\pi/2) = 1$.

3. Let Δt denote a small change in t. The difference of a function $x(t)$ over the interval $(t, t + \Delta t)$ is then denoted by $\Delta x = x(t + \Delta t) - x(t)$. The average rate of change of $x(t)$ over that interval is $\Delta x / \Delta t$. In the limit $\Delta t \to 0$ one obtains the (instantaneous) rate of change denoted by the calculus notation $\dfrac{dx}{dt}$ ('derivative of x with respect to t').

   If $x(t)$ is the displacement of a particle with time, then $v = \dfrac{dx}{dt}$ is its velocity and $a = \dfrac{dv}{dt}$ the acceleration. We may write $a = \dfrac{d^2 x}{dt^2}$ ('second derivative of x with respect to t').

4. For a function $f(x, t)$ of two independent variables x and t, we may find the rate of change of f with respect to one variable while keeping the other constant. For example, keeping x constant we denote the rate of change with respect to t by the *partial derivative* $\dfrac{\partial f}{\partial t}$.

5. A vector $\mathbf{v}$ may be written in terms of its components in Cartesian coordinates, $\mathbf{v} = v_1 \mathbf{i} + v_2 \mathbf{j} + v_3 \mathbf{k}$ where $\mathbf{i}$, $\mathbf{j}$, $\mathbf{k}$ are the constant unit vectors along the three axes. The scalar (dot) product of two vectors is then

$$\mathbf{u} \cdot \mathbf{v} = u_1 v_1 + u_2 v_2 + u_3 v_3 . \tag{2.4}$$

The vector (cross) product is

$$\mathbf{u} \times \mathbf{v} = (u_2 v_3 - u_3 v_2)\mathbf{i} + (u_3 v_1 - u_1 v_3)\mathbf{j} + (u_1 v_2 - u_2 v_1)\mathbf{k} . \tag{2.5}$$

The vector differential operator[8] ∇ ('Del') is

$$\nabla = \mathbf{i}\frac{\partial}{\partial x} + \mathbf{j}\frac{\partial}{\partial y} + \mathbf{k}\frac{\partial}{\partial z} . \tag{2.6}$$

---

[8] An 'operator' acts on a function, as for example in Eq.(2.7).

So, for example, acting on a vector function $\mathbf{V}$, with components V_x etc., we have

$$\nabla \cdot V = \frac{\partial V_x}{\partial x} + \frac{\partial V_y}{\partial y} + \frac{\partial V_z}{\partial z} \ . \tag{2.7}$$

The *Laplacian* ∇^2 is the differential operator

$$\nabla^2 \equiv \nabla \cdot \nabla = \frac{\partial^2}{\partial x^2} + \frac{\partial^2}{\partial y^2} + \frac{\partial^2}{\partial z^2} \ . \tag{2.8}$$

6. A *Group G* is a set equipped with an operation $\circ$ which satisfies these properties: for any two elements a and b in G, $a \circ b$ is also in G (*closure* property); the operation is *associative*, $a \circ (b \circ c) = (a \circ b) \circ c$; there exists a unique *identity* element e satisfying $e \circ a = a \circ e = a$ for all a in G; for each a in G there is a b in G (the *inverse*) such that $a \circ b = b \circ a = e$.

   Groups are used in the mathematical study of symmetries. If $a \circ b = b \circ a$ for all elements of G then the group is *Abelian* (*commutative*). Most of the important groups in physics, such as the three-dimensional rotation group, are *non-Abelian*.

2.9   Exercises

1. Your mobile phone screen has gone blank. You press the power button, but nothing happens.

   (a) List the possible reasons for this phenomenon and discuss how the scientific method may be used to solve the problem. Consider also the role of Occam's Razor in prioritising the various hypotheses.

   (b) Discuss another example from daily life that illustrates the scientific method.

2. You are asked to investigate the unknown relationship between an independent variable x, and a variable y that is believed to be dependent on x. You are given two pairs of data points, (x_1, y_1) and (x_2, y_2).

   (a) Would that be enough to determine the function $y(x)$?

   (b) If not, how many points would you need to create a model for $y(x)$?

   (c) Could you create more than one model even with the number of points you chose in (b)?

   (d) How could you narrow down the possibilities and choose the 'most plausible' or 'realistic model' from among a few alternatives?

   (e) What is the difference between *correlation* and *causation*? How could you differentiate between those two possibilities?

3. What is the difference between a sceptic and a cynic? Is it possible to be sceptical of everything in the world and our interactions with it?

4. Explain, with examples, the differences between science, pseudo-science, non-science, and nonsense.

5. Debate one of these topics with your colleagues:

   (a) There are many things we do not yet understand. Physicists should, therefore, keep an open mind to phenomena such as extra-sensory perception (ESP) and astrology.

   (b) Physics is a social construct like all other forms of human activity; it cannot claim to describe an objective reality, especially since no one knows what reality is. Hence we should put no more faith in physics than in other opinions and philosophies.

6. Estimate the number of atoms in the air in your room.

7. You might have heard the expression "the human body is mostly water". Use that fact to estimate the masses of your heart and brain. Compare your estimates with tabulated values.

8. Neo guesses that the period (T) of a satellite in circular orbit around the Earth depends on the gravitational constant (G), the distance of the satellite from the centre of the Earth (R), and the masses of the Earth (M) and satellite (m).

   (a) Determine the dimensions of G from Eq.(4.13).

(b) Use dimensional analysis to obtain a possible relation giving T in terms of the other variables.

(c) Compare your expression with Kepler's law, $T^2 \propto R^3$.

9. Even when an object moves in a circle at constant speed, its velocity changes because of the change in its direction of motion. Hence the object experiences an acceleration called *centripetal acceleration.*

(a) Use dimensional analysis to deduce how the centripetal acceleration a, depends on the mass of the object m, its speed v, and the radius of the circle r.

(b) Show that as $r \to \infty$ your expression for a tends to zero. Why is this expected?

(c) Show that as $v \to 0$ your expression for a tends to zero. Why is this expected?

(d) Can you deduce the direction of the acceleration using symmetry arguments, or otherwise?

10. Neo thinks that the height h, reached by a cannon ball fired vertically upwards with an initial speed v from the surface of the Moon is given by $h = k/v^2$, where k is some unknown constant.

(a) If the equation were correct, what would be the dimension of k?

(b) Could Neo's equation for h be correct?

11. Show that the following are groups:

(a) The set of integers, under the operation of addition.

(b) The set consisting of the two elements -1 and 1, under the operation of multiplication.

(c) The set of vectors in the plane, under the operation of vector addition.

(d) The set of all 2×2 matrices with real entries and non-zero determinant, under the operation of matrix multiplication.

Your Notes

Michael Faraday
1791-1867

He made several important contributions to the study of
electromagnetism, including the discovery of electromagnetic
induction. His idea of 'lines of flux' eventually led to the
development of field theory.

3

Space, Time and Motion

3.1 Space and Time

Initial evidence suggested, and we assume throughout this book, that the *space* in which we are embedded, and in which we observe events, is *Euclidean*. Thus various fixed properties hold: sums of angles of a triangle add up to 180 degrees, the ratio of the circumference of a circle to its diameter equals π, and Pythagoras' theorem relates the lengths a, b, c of a right-angled triangle, $a^2 + b^2 = c^2$ with c the hypotenuse. The properties of this space are unaffected by the presence of matter or energy and so, in this sense, space is *absolute* [4].

In Newtonian mechanics, it is also assumed that there exists an observer-independent quantity called *time* that is unaffected by the presence of matter or energy [4]. The passage of this absolute time can then be used to track the various changes we observe. In practice, however, we construct clocks to indicate the passage of time through the change in relative position of some quantity, such as the shadow of a sundial, or the oscillation of a crystal [6]. In essence, then, time is what a clock measures, and we assume in the following that reliable, synchronised clocks are available.

3.2 Matter

Ordinary matter [7] is composed of different types of *atoms*. At the classical level, as in most of this book, atoms can simply be modelled as hard billiard balls with a diameter of about 10^{-10} m. When a more detailed model is required, an atom can be modelled as a small *nucleus*, of diameter about 10^{-15} m, surrounded by *electrons* which are point-like. The nucleus contains *protons* and *neutrons*, all of which are tightly bound.

Atoms of different type are distinguished by their *atomic number*, which is the number of protons in the nucleus. A *molecule* is a bound combination of atoms. For example, a molecule of water has two hydrogen atoms and one oxygen atom.

Macroscopic pieces of matter contain a large number of atoms, of the order of Avogadro's number, $\sim 10^{23}$. However, even when dealing with macroscopic pieces of matter, it is often convenient to first idealise them as point-like objects which we will refer to as *particles*. The discussion of extended objects is taken up in Sect.(4.2).

3.3   Kinematics

Kinematics is the study of motion, without enquiring about the causes of motion. To describe the motion of a particle through space, we need to quantify both the passage of time using clocks, and the location of the object.

The location of a particle is referenced with respect to some *frame*. For example, an aeroplane may be tracked by an observer in a control tower, which would be one frame of reference, and the aeroplane may also be monitored by another observer in a moving car, which would define a second frame of reference.

If the Cartesian coordinate system is chosen in a frame, the position of the particle is indicated by its time-dependent coordinates $\mathbf{r} = (x(t),\ y(t),\ z(t))$. We may also compile the information in the form of four coordinates $(t,\ x,\ y,\ z)$. The four coordinate form is convenient for specifying an *event*: an occurrence at a specific location and time.

It is important to note that *coordinate systems* are simply mathematical constructs we use to help us in our investigations, and we are free to choose any convenient coordinate system. Physical phenomena are independent of the coordinate system selected by a particular observer, though their mathematical description may appear different (for example, when describing the flight of a plane in Cartesian or spherical coordinates).

Suppose that a particle, initially at a location indicated by the position vector $\mathbf{r}$, changes its location slightly. The change in its position, called *displacement*, is denoted by $\Delta \mathbf{r}$, where we use the symbol Δ ('delta') to indicate a small change; so $\Delta \mathbf{r}$ means 'a small change in $\mathbf{r}$'. The *velocity* of a particle is defined as the rate of change of its displacement with time, and tentatively we write $\mathbf{v} \sim \frac{\Delta \mathbf{r}}{\Delta t}$. Since objects generally do not move at constant velocity, we take the limit $\Delta t \to 0$ of that ratio to obtain the (instantaneous) velocity, and use the calculus notation, $\mathbf{v} = \frac{d\mathbf{r}}{dt}$. The components of the velocity along the three axes are denoted by v_x, v_y and v_z.

Velocity is a vector quantity; it defines not only speed but also a direction. For example, 'constant velocity' means not only constant speed, but also a fixed direction: a ball tied to the end of a string can be made to travel at a constant speed around a circle, but since its direction of motion is changing its velocity is not constant.

Acceleration is the rate of change of velocity with time, $\mathbf{a} = \frac{d\mathbf{v}}{dt}$. So acceleration too is a vector.

3.4   Inertial Frames

Suppose we have a frame of reference, S, that is at 'rest'. At rest relative to what? For some applications, a frame attached to the earth would be a suitable stationary frame, while for other applications, as discussed below, a frame attached to the Sun or distant stars would be a better choice.

Let S' be another frame moving at constant velocity relative to S. Galileo conducted experiments which convinced him that observers attached to each of those frames would experience the same laws of physics. In other words, one could not perform mechanical experiments within such frames to discover one's motion (velocity). This is *Galileo's Principle of Relativity*, which was later generalised by Poincaré and Einstein to all physical phenomena, including, for example, electromagnetism.

Those special frames of reference, each moving at constant velocity, are called *inertial frames* [8]. (Of course, from the point of view of any inertial frame, it is at rest, while the other inertial frames are moving relative to it).

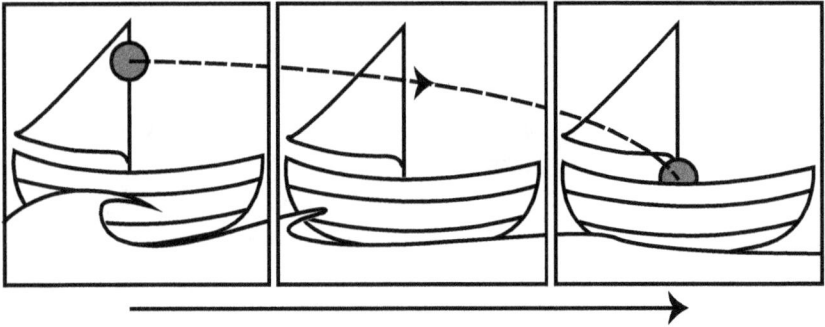

Figure 3.1: In a uniformly moving ship, an object released from the top of the mast ends at the bottom of the mast, just as in a stationary boat.

3.5 Galilean Transformations

Consider two inertial frames. The first, referred to as the S frame, is what we will typically think of as the 'stationary frame'. It will have the Cartesian coordinate system (x, y, z) with an observer (Stan) located at the origin O and a clock that measures time t. The second frame, S', has coordinate system (x', y', z') and moves with constant velocity u relative to the S frame in the x-direction, with its y' and z' axes parallel to the y and z axes respectively, and its observer (Maria) is located at O', see Fig.(3.2). We further assume that the clocks in S and S' are synchronised and that $t = t' = 0$ when the two origins O and O' coincide.

Consider an event as described from those two frames. Since time is an absolute quantity in Newtonian physics, we will always have $t = t'$. Let the location of the event have coordinates (x, y, z) in the S frame. Then its coordinates in the S' frame will satisfy $y' = y$ and $z' = z$, but since S' moves to the right we will have $x' = x - ut$. The set of equations

$$t' = t ; \tag{3.1}$$
$$x' = x - ut ; \tag{3.2}$$
$$y' = y ; \tag{3.3}$$
$$z' = z , \tag{3.4}$$

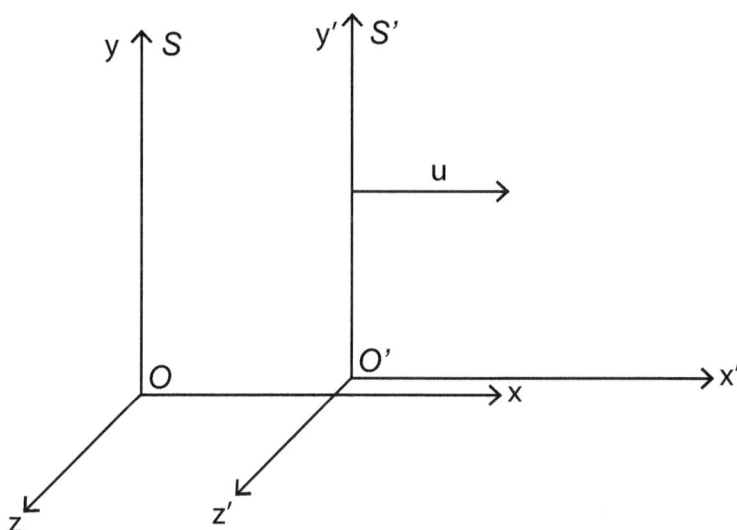

Figure 3.2: Two inertial frames in relative motion.

are referred to as the *Galilean transformations*. They describe the same event as observed from two different inertial frames.

Consider now a particle moving from point 1 to point 2. Then[1] from (3.2) we have $x_1' = x_1 - ut_1$ and $x_2' = x_2 - ut_2$. Subtracting those equations from each other gives $\Delta x' = \Delta x - u\Delta t$ and then dividing by Δt gives

$$v_x' = v_x - u \tag{3.5}$$

and similarly $v_y' = v_y$, $v_z' = v_z$. These expressions relate the velocities of a particle as observed from the two different inertial frames.

3.6   Invariants

A powerful idea in physics is that of *invariants*, that is, quantities that do not change in the course of our investigations. Of course, when talking about change, we must specify the context. Here we focus on quantities that do not change their value when we shift observations from one inertial frame to another.

---

[1] x_1 denotes the x-coordinate of the particle at point 1, as measured in the S frame, etc. We also write $x_2 - x_1 = \Delta x$ etc.

Quantities that have the same value when measured from different inertial frames related by Galilean transformations are called *Galilean invariant*. Which quantities remain unchanged under Galilean transformations? Since Newtonian time is absolute, it is not surprising that time intervals are Galilean invariant. Mathematically, if t_1 and t_2 are the times of two events as measured in the S frame, then $\Delta t = \Delta t'$ where the prime refers to quantities in the S' frame.

What about lengths? Suppose a rod is at rest along the x'-axis in the S' frame with ends at x_1' and x_2', so that its length in the S' frame is $\Delta x'$. If Stan, in the S frame, were to measure the length of the rod, he would need to determine the location of its two ends at the same time in his frame. Thus from (3.2) we get $\Delta x = \Delta x'$. So lengths are Galilean invariant.

Similarly, although velocities are relative, as we see from Eq.(3.5), a particle's acceleration is the same in all inertial frames (see exercises).

The inertial mass of an object, introduced through Newton's laws in the next chapter, is Galilean invariant by definition.

3.7 Exercises

1. Maria is seated in a car moving at constant velocity.

    (a) If she held her mobile phone at head level and released it, what would the path it takes to her lap look like from her perspective? What would the path look like to Stan who is stationary and observing from the road?

    (b) If Maria threw her phone vertically up, along which trajectory would it fall back?

    (c) What would happen in part (a) if the car accelerated?

    (d) If you are enclosed in a car moving in a straight line, what simple experiment could you perform to deduce whether you were accelerating?

2. Consider the Galilean transformation of coordinates between two inertial frames moving relative to each other along the x-direction as in Eqs.(3.1-3.4).

(a) Derive the transformation rule for the acceleration of a particle.

(b) Two events are *simultaneous* if they occur at the same time. Is simultaneity a Galilean invariant notion? That is, if $t_1 = t_2$, is it then always true that $t'_1 = t'_2$ regardless of the values of x_1 and x_2?

(c) Is the notion of 'at the same place' Galilean invariant? Is this surprising?

3. Derive the inverse Galilean transformation corresponding to Eqs.(3.1-3.4). That is, express (t', x', y', z') in terms of (t, x, y, z), in two different ways:

(a) By direct algebraic manipulation of those equations.

(b) By noting that from the perspective of S', it is the frame S that is moving at velocity $-u$. So the relation giving the primed variables in terms of the unprimed variables should simply be obtained by interchanging the primed and unprimed symbols in the equations, together with the replacement $u \rightarrow -u$.

4. Suppose that the coordinate axes of the two inertial frames in Fig.(3.2) were not aligned as shown. How would the subsequent discussions and conclusions change?

5. Why do raindrops hitting the side windows of a moving car often leave diagonal tracks?

6. A 'moving sidewalk' in an airport moves at 2 m/s and is 25 m long. A woman steps on at one end and walks at 2.5 m/s relative to the sidewalk. Determine how much time she would take to reach the opposite end if she walked

(a) in the same direction as the movement of the sidewalk.

(b) in the opposite direction to the movement of the sidewalk.

7. Points A and B are located a distance 1200 m apart along the banks of a straight river which is flowing at a constant velocity of 1 km/h. How much time would Plato take to row from A to B and back, if he rows at a constant speed of 2 km/h relative to the water?

8. A particle is moving at constant speed v around a circle of radius r. Define its *angular speed* ω as the rate of change of its angular position (measured in radians). Show that $v = r\omega$.

9. A particle is moving at an unknown non-zero speed around a circle of radius r. What can you say about the magnitude and direction of its acceleration?

10. Why does time seem to go only in one direction? Are there different arrow's of time, such as thermodynamic, cosmological, biological and psychological?

11. According to Aristotle, a force is required to keep an object moving at constant velocity. If true, then

   (a) What is the state of a particle when no forces act on it?
   (b) What would the above state appear to a person in a relatively moving frame?
   (c) How does this compare with Galileo's principle of relativity?

12. Consider the horizontal plane with Cartesian $x - y$ axes and centre at O. Let $R(\theta)$ denote a rotation of the axes in the plane about O by an angle θ; and let $T(\mathbf{a})$ denote a translation (displacement) of the coordinate system in the plane by the vector $\mathbf{a}$.

   (a) Show that the set consisting of $R(\theta)$ for all θ forms a group.
   (b) Show that the set consisting of $T(\mathbf{a})$ for all $\mathbf{a}$ forms a group.

   Note: These are examples of 'continuous groups' since the parameter labelling the group elements is continuous.

13. Let the *Galilean boost* $G(u)$ represent the operation described by Eqs.(3.1-3.4) whereby one inertial frame is transformed to another. Show that the set consisting of $G(u)$ for all u forms a group.

14. If all motion is relative, then why is the heliocentric model of the solar system preferred over the geocentric (or 'Marscentric') model?

4

Classical Mechanics

In contrast to Aristotle's view, Galileo argued, through thought and real experiments, that it was not necessary to apply a force to keep an object moving at constant velocity. His *law of inertia* formed the basis for Newton's First Law of motion.

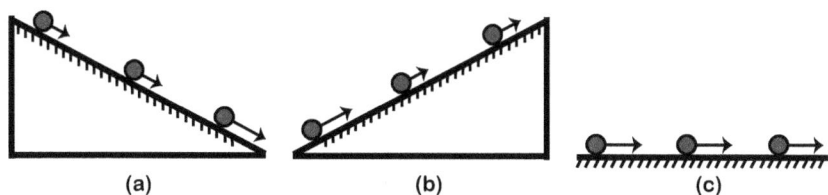

Figure 4.1: A ball rolling down a slope accelerates, while one rolling up decelerates. Thus, Galileo argued, in the limiting case, a ball rolling on flat ground would maintain a constant velocity.

4.1 Newton's Laws

Newton formulated three laws of motion:

1. *First Law*: In an *inertial frame*, a particle remains at rest, or continues its motion at constant velocity, unless acted on by a *force*.

   If we have a prior understanding of what an 'inertial frame' means, as in the last chapter, then the First Law can be taken to define a *force* as an interaction between the particle and its surroundings which tends to cause a change in its state of motion.

On the other hand, the First Law may be turned around to identify non-inertial frames: if you cannot find any interaction between a particle and its surroundings (that is, no 'forces' act on the particle), and yet you observe a change in the velocity of the particle, then you are not in an inertial frame [9].

2. Define first the *inertial mass* of a particle as quantifying its matter content. We discuss in Exercise (6) of the next chapter how such a quantity can be measured. Next, define the (linear) *momentum* of the particle as

$$\mathbf{p} = m\mathbf{v} \qquad (4.1)$$

where $\mathbf{v}$ is its velocity.

Second Law: In an inertial frame, the force $\mathbf{F}$ acting on the particle equals its rate of change of momentum,

$$\mathbf{F} = \frac{d\mathbf{p}}{dt} \ . \qquad (4.2)$$

This general form of the law applies even if the system mass changes. If the mass is constant, then we may also write the law as

$$\mathbf{F} = m\mathbf{a} \qquad (4.3)$$

where $\mathbf{a}$ is the acceleration.

Newton's Second Law relates a 'cause' (the force) with a certain 'effect' (the change in momentum). For it to be useful, we need explicit expressions for the forces [10]. We will discuss two of the most important forces[1] in Sects.(4.4) and (4.5).

The inertial mass appears as a proportionality constant in Eq.(4.3) which explains its name: it is a measure of the object's inertia or resistance to change. We will simply refer to the inertial mass as *mass*.

3. *Third Law*: If one particle exerts a force on a second particle, then the second particle exerts an equal but opposite force on the first particle.

---

[1]See also Eq.(8.3).

4.2 Centre of Mass

If the system of interest consists of a collection of particles of masses m_i, with $i = 1, ..., N$, we can add up the individual forces $\mathbf{F}_i$ acting on the particles to get the total force $\mathbf{F}$ on the system,

$$\mathbf{F} = \sum_{i=1}^{N} \mathbf{F}_i = \sum_{i=1}^{N} m_i \mathbf{a}_i , \qquad (4.4)$$

where

$$\mathbf{a}_i = \frac{d^2 \mathbf{r}_i}{dt^2} \qquad (4.5)$$

and $\mathbf{r}_i$ is the position vector of the i-th mass relative to an origin of an inertial frame. $\mathbf{F}$ is actually the total external force acting on the system since the internal forces appear in pairs and cancel due to Newton's Third Law.

If the system is an extended object, we imagine dividing it into N smaller pieces and use the same definitions as here[2]. The *centre of mass*, $\mathbf{r}_{cm}$, of a system of total mass M is defined by the weighted average

$$\mathbf{r}_{cm} = \frac{\sum_i m_i \mathbf{r_i}}{\sum_i m_i} = \frac{\sum_i m_i \mathbf{r}_i}{M} . \qquad (4.6)$$

We deduce that the acceleration of the centre of mass, $\mathbf{a}_{cm}$, is given by

$$\mathbf{a}_{cm} = \frac{d^2 \mathbf{r}_{cm}}{dt^2} = \frac{\mathbf{F}}{M} . \qquad (4.7)$$

That is, the motion of the centre of mass follows Newton's Second Law: regardless of the complicated dynamics of parts of the system, the motion of the centre of mass is relatively easy to describe.

4.2.1 Momentum

The total (linear) momentum $\mathbf{P}$ of the system coincides with the momentum of the centre of mass:

$$\mathbf{P} = \sum_i m_i \frac{d \mathbf{r}_i}{dt} = \frac{d}{dt} \sum_i m_i \mathbf{r}_i = M \mathbf{v}_{cm} . \qquad (4.8)$$

---

[2]The continuum limit is taken by replacing sums with integrals.

So we may re-write Eq.(4.7) as

$$\mathbf{F} = \frac{d\mathbf{P}}{dt} .$$

(4.9)

4.2.2 Angular Momentum

Fig.(4.2) shows two identical particles, each located at one end of a light uniform rod, with equal but opposite forces acting on them. As the net force on the system is zero, there will be no motion of the centre of mass if it is initially at rest. Yet, the rod will clearly rotate about its centre.

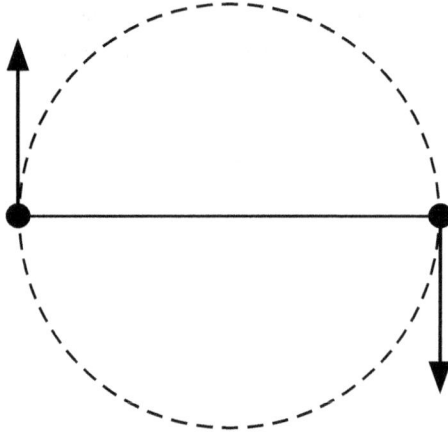

Figure 4.2: Two identical particles at the ends of a light rod. Forces of equal magnitude are applied in the direction indicated by the arrows.

The rotational effect of forces on a system is measured by a quantity called the *torque*. It is defined by

$$\mathbf{N} = \sum_i (\mathbf{r}_i \times \mathbf{F}_i)$$

(4.10)

where i indexes the individual masses and, as before, position vectors are measured relative to the origin of an inertial frame[3].

---

[3]If the centre of mass does not move, then it can be conveniently chosen as the origin of a 'rest' frame.

Define the *angular momentum*, **L**, of a system by

$$\mathbf{L} = \sum_i (\mathbf{r}_i \times \mathbf{p}_i) \tag{4.11}$$

where $\mathbf{p}_i$ is the linear momentum of the *i*-th mass. Then Newton's laws give the result

$$\mathbf{N} = \frac{d\mathbf{L}}{dt}, \tag{4.12}$$

which is the rotational analogue of Eq.(4.9) [12].

4.3   Fundamental and Emergent Interactions

Four fundamental interactions have been identified: gravity, the electromagnetic force, the strong force, and the weak force.

Gravity, as described by Newton, is discussed in the next section [13].

The electromagnetic interaction is the *unified* description of electric and magnetic effects, and is discussed in Chapter 8.

The *strong force* is responsible for keeping the protons and neutrons confined within the nucleus of an atom. It is stronger than the electromagnetic force but has a shorter range: its effects do not manifest far beyond the nucleus.

The *weak force* is extremely weak, and short ranged, but it is responsible for radioactive decay. The electromagnetic and the weak interactions have been unified into the *electroweak* theory, and there are ongoing attempts to extend this unification to the rest of the fundamental forces.

What about other forces that one encounters, like friction, Hooke's law, or 'hydrogen-bonds'? Those are *emergent* forces[4], whose microscopic origin is the electromagnetic force.

On the other hand, what we view as 'fundamental forces' today might well be understood to be emergent tomorrow.

4.4   Newton's Law of Gravitation

Newton's Law of Universal Gravitation states that the force of attraction on a particle of mass m by another particle of mass M is

---

[4]See Sect.(7.7) for the definition of emergence.

directly proportional to their masses, and inversely proportional to the square of the distance between them:

$$\mathbf{F} = -\frac{GMm}{r^2}\,\hat{\mathbf{r}}\,, \qquad (4.13)$$

where $G = 6.67 \times 10^{-11}$ N m^2/kg^2 is the gravitational constant, and $\hat{\mathbf{r}}$ is a unit vector pointing from mass M to mass m [14]. The law (4.13) also applies to uniform spherical masses, with r the distance between the centres of the masses.

If a particle of mass m is released from rest above the Earth, approximated as a uniform sphere of mass M, then Newton's Second Law together with the law of gravity (4.13) imply that the acceleration of the particle is $g \equiv \frac{GM}{r^2}$. The acceleration is independent of the particle's mass, as Galileo found, but depends on r. Near the surface of the Earth, one can put $r \approx R_E$, the radius of the earth, and get the typical value for g of about 9.8 m/s^2. The *weight* of a particle is the force it experiences due to gravity, which would be mg.

On the other hand, if the released particle has a sufficiently large tangential velocity, then Newton deduced from his equations that it would enter into an elliptical orbit, in agreement with the earlier empirical findings of Kepler.

4.5   Hooke's Law and Harmonic Motion

Consider a single particle at rest while under the action of various forces. It is then at an *equilibrium* point [15]. The equilibrium point is said to be *stable* if the particle experiences a restoring force when it is displaced slightly from that point. If the displacement of the particle is restricted to a single direction labelled by x, then the restoring force for small x would be [16]

$$F(x) = -kx \qquad (4.14)$$

where $k > 0$ is a constant whose value depends on the parameters of the system. The relation (4.14) is called *Hooke's Law* . The equation of motion of the displaced particle is obtained by combining Eq.(4.14) together with Eq.(4.3), giving [17]

$$m\frac{d^2x}{dt^2} + kx = 0\,. \qquad (4.15)$$

The general solution of Eq.(4.15) can be shown to be

$$x(t) = A\sin(\omega t + \phi) \tag{4.16}$$

where A is the amplitude of oscillations and ϕ is a constant; their values are determined by the initial ($t = 0$) conditions of the system. The period of oscillations T is related to the *angular frequency* ω by $\omega = \sqrt{k/m} \equiv 2\pi/T$.

The perfectly sinusoidal solution (4.16) is called *simple harmonic motion* [18]. It can be approximately realised by a simple pendulum, which consists of a small heavy bob attached to the end of a light string of length l whose other end is fixed. The small angle oscillations of the bob in the vertical plane have a period given by $T = 2\pi\sqrt{l/g}$ where g is the acceleration due to gravity [19].

4.5.1 Resonance

If an external periodic force is applied to an oscillating system, such as a swinging pendulum, the amplitude of oscillations increases greatly when the period of the external force matches the natural period of the system. This phenomenon is called *resonance*.

4.6 Exercises

1. An ox, which has heard of Newton's Third Law, but not understood it very well, refuses to pull a cart attached to it because it argues: "If I pull on the cart with a certain force, Newton's Third Law will imply an equal but opposite force, which would then mean that the total force on the cart is zero, and so the cart will not move. So why waste my effort? I won't pull". What's wrong with the ox's argument?

2. Maria is sitting in a car which makes a sharp right turn.

   (a) In which direction would she tend to move? Would you say that a force was acting on her?

   (b) How would the situation look to Stan who was stationary and observing the event from the road?

   (c) How are the analyses in parts (a) and (b) consistent with Newton's laws?

    (d) Why does the top part of Maria's body move more than the bottom part?

3. You are initially standing still. Then you walk. Using Newton's laws of motion, describe how the change from the stationary state to the walking state takes place.

4. Twin brothers Akshat and Ankit have identical body shapes, weight and strength, and are equally proficient in karate. While Akshat remains still and passive, Ankit hits his brother with a powerful karate chop. A student who observes the incident argues: "That was pointless because by hitting Akshat, Ankit would experience an equal and opposite force on himself and so suffer just as much". Do you agree? Explain.

5. A car is moving at constant speed along a straight road. Draw a diagram to indicate all the forces acting on the car and explain the origin of those forces. (Do not confuse forces acting on the car with the forces acting on the road).

    (a) Is friction between the tires and the road good or bad?

    (b) What determines the force of friction between the road and the tires?

6. Describe two ways in which you could make yourself seem (from the scale reading) lighter when standing on a weighing scale. How could you make yourself seem heavier?

7. A mountain made of rock of density ρ exerts a pressure $\rho g H$ on its base, where H is the height of the mountain and g is the acceleration due to gravity.

    (a) Estimate the maximum height of mountains on Earth using the limiting value of pressure that the strongest rock can withstand.

    (b) How are mountains formed? What decreases their height?

    (c) Estimate the height of mountains on Mars.

    (d) Compare your estimates above with the actual heights of mountains on Earth and Mars.

8. A satellite is launched into a circular orbit above the Earth's equator. An observer on the ground notices that the position of the satellite does not change. Calculate the height of the satellite above ground.

9. Why is it that you sometimes feel "light" or "heavy" while in a fast moving lift?

10. If gravity is a universally attractive force, then why do objects have a shape and size? For example, why don't the different parts of your body continue to attract each other until your whole mass is concentrated into a point (a *black hole*)?

11. Which is easier to stop with your bare hands, and why: a bicycle, or a car, both travelling at 1 m/s.

12. Where in a car is the 'crumple zone' located? What role does it play, and what are the principles on which it operates?

13. Sand falls at a constant rate from a vertical funnel onto a horizontally moving conveyor belt. Would a force be required to keep the belt moving at constant speed? (Ignore frictional losses in the gear mechanism).

14. What is the *centre of weight* of an object? Does it always coincide with the centre of mass?

15. Discuss one example where resonance is encouraged, and another example where it is to be avoided.

16. Show that Eq.(4.3) keeps the same form (*form invariance*) under Galilean transformations if the expression for the force is the same in both frames, $\mathbf{F}' = \mathbf{F}$. Verify that Newton's law of gravitation and Hooke's law satisfy that condition.

17. In non-inertial frames, such as a rotating platform, the deviation of an object's trajectory from that expected in inertial frames can be described using 'pseudo-forces' such as the 'centrifugal force' and the 'Coriolis force'. Illustrate the action of those pseudo-forces with explicit examples.

Your Notes

Emmy Noether
1882-1935

Her key contribution to physics is the influential Noether's
Theorem, in which she established the connection between
conservation laws and the symmetries of a system.

5

Conservation Laws and Symmetries

Under certain conditions, even highly complex systems have associated quantities whose values do not change with time. The existence of such conserved quantities allows us to place constraints on the possible dynamics of a system even though we might be ignorant of various details. This is particularly important in the case of systems with many components, where it might be too difficult to solve for and follow in detail the trajectory of each part as described by Newton's laws of motion.

Noether's Theorem [10] states that the continuous[1] symmetries of a physical system give rise to conservation laws. By *symmetry* we mean that something, geometrical or abstract, remains *invariant* (does not change its form) when we carry out an operation on the system, such as moving it to the right or rotating it.

5.1 Conservation of Energy

If the laws of physics describing a system do not change with time, then there is a conserved quantity called *energy* associated with the system. The total energy in a closed system is a constant though it may manifest in different forms[2]. The Law of Energy Conservation is one of the most important laws of nature.

In Newtonian mechanics, the *kinetic energy* of a particle of mass m travelling with velocity $\mathbf{v}$ is given by $T = \dfrac{mv^2}{2}$.

---

[1]Strictly speaking, one requires differentiable (smooth) variables.

[2]In an open system, energy can increase or decrease. Energy would be conserved though in a larger closed system that contains the original open system.

Energy due to position or the internal configuration of a material body is called *potential energy*. For example, the energy stored in a compressed spring is called elastic potential energy. A stone on a mountain top has gravitational potential energy. A charged particle stationary in an electric field has electrostatic potential energy.

The *gravitational potential energy* of a mass m in a uniform gravitational field with acceleration g is given by $U_g = mgh$, where h is the height above the chosen reference point.

The *work*, W, done by a force $\mathbf{F}$ in displacing a particle by an infinitesimal amount $\Delta\mathbf{r}$ is defined by[3]

$$W = \mathbf{F} \cdot \Delta\mathbf{r} . \tag{5.1}$$

Work has the same units as energy. The work done on or by an object changes its energy.

5.2 Conservation of Momentum

We see from Eq.(4.9) that if the total external force acting on a system is zero, then the total (linear) momentum does not change with time: it is conserved.

The conservation of linear momentum is associated with the symmetry of the system under linear displacements (*translations*). That is, the system is described by the same physics equations when the whole system is shifted linearly in space.

As momentum is a vector, it is possible for momentum to be conserved in some directions (along which no external forces act), even if it is not conserved in other directions.

The principle of momentum conservation explains how rocket propulsion is possible even in empty space. The momentum of fast moving gases expelled from the rear of the rocket is balanced by the creation of a forward momentum for the rocket.

5.3 Conservation of Angular Momentum

We see from Eq.(4.12) that if the total external torque acting on a system is zero, then the total angular momentum is conserved. The

---

[3]For a finite displacement, the work done is given by a line integral $\int_C \mathbf{F} \cdot d\mathbf{r}$.

conservation of angular momentum is associated with the symmetry of the system under rotations.

Just as in the case of linear momentum, it is possible for angular momentum to be conserved in some directions (along which no external torques act), even if it is not conserved in other directions.

Angular momentum conservation allows an ice-skater to increase her rotational speed by drawing in her arms as she spins.

5.4 Conservation of Mass

In Newtonian physics, the total mass of a system is simply the sum of the masses of its constituents, and that sum is conserved in many circumstances.

The exceptions are noticeable for some processes, such as those at very high energies or involving radioactivity. For example, in the Sun two hydrogen atoms fuse to form a helium atom, releasing large amounts of energy. The mass of a helium atom is less than the mass of the two hydrogen atoms.

As the 'lost' mass is tiny even when large amounts of energy are released, we can take it that the total mass of a system, defined in Newtonian physics as the sum of the masses of its components, is conserved to a good approximation.

It is possible to define a system mass M that is conserved exactly, but it is not merely the sum of the masses of the individual components [20].

5.5 Conservation of Charge

Just like mass, electric *charge* is a property of matter. A particle can be positively charged, like the proton; negatively charged, like the electron; or neutral, like the neutron. The charge of a particle determines its interaction with electromagnetic fields, see Chap.8.

The total electric charge in a closed system is conserved. This conservation law is a consequence of Maxwell's equations [38] and the associated symmetry is called *gauge invariance* [21].

Unlike the space-time symmetries associated with energy, momentum and angular momentum conservation, gauge invariance is an example of an *internal symmetry*. Identifying internal symmetries is

the usual approach taken for understanding the fundamental particles and their interactions[4].

5.6 Exercises

1. A ping-pong ball is released from a height of 3 m onto a hard
   surface.

   (a) What height will the ball rebound to and why?

   (b) Describe all the energy conversions that take place from
       the moment the ball is released to the time it finally comes
       to rest.

   (c) Repeat the above analyses for a ball released on the Moon.

2. Estimate how high you would have to climb the stairs of a building to expend the energy you gain after consuming 100 grammes
   of your favourite chocolate bar. (Is it worth it?)

3. You find yourself stranded on a boat in the middle of an ocean.
   There are no sea currents, no waves and no wind. There is
   some equipment in the boat: a rope, a canvas cloth, a long
   pole, a powerful electric fan, and an electric power supply to
   run the fan. Suggest an efficient way to move the boat forward,
   explaining the physical principle(s) involved with the help of a
   sketch. (The fan is not waterproof; it cannot function under
   water).

4. Consider Fig.(4.2) with the forces set to zero. Suppose that
   the two masses have the same angular speed ω, and the rod
   is of negligible mass and length $2r$. Show that the angular
   momentum of the system about the centre is $L = 2mr^2\omega$. Use
   this to explain how a ballerina can increase her rotational speed
   while spinning.

5. Most helicopters have a single large rotating rotor which pro-
   vides the lift force. Explain how the lift force is generated, and
   why the body of the helicopter does not spin in the opposite direction to the rotor's spin (as angular momentum conservation
   would suggest).

---

[4]See Sect.(4.3).

6. A particle of mass M is stationary in an inertial frame. Another particle of mass m is sent on a head-on collision towards the first mass at a velocity u along the x-axis. Show that before the collision the momentum of the system is $P = mu$. If the velocities of m and M immediately after the collision are v and w respectively (along the x-axis), show that $m = \dfrac{Mw}{u - v}$. Use this to explain how masses might be calibrated in principle.

7. Is it possible for a quantity to be conserved, but not be Galilean invariant?

8. A bicycle wheel is suspended from one end of its axle and held with the axle horizontal. In that position, the wheel is set spinning by hand and then released. Describe what happens next. Does this have an application?

9. Create a simple physics based model to determine the maximum height a pole-vaulter may reach. Compare with real world data and other models in the literature.

10. Two particles of masses m_1 and m_2 move along the x-axis of the S frame with velocities v_1 and v_2 respectively. They collide, combine for a very short while, and then break up into two pieces of masses m_3 and m_4 that move along the same axis with velocities v_3 and v_4 respectively.

   (a) Show that the law of conservation of momentum implies

   $$m_1 v_1 + m_2 v_2 - m_3 v_3 - m_4 v_4 = 0. \qquad (5.2)$$

   (b) Rewrite Eq.(5.2) in terms of velocities as measured by an observer in the S' frame, see Eq.(3.5), to get

   $$m_1 v_1' + m_2 v_2' - m_3 v_3' - m_4 v_4' = (m_3 + m_4 - m_2 - m_1)u. \qquad (5.3)$$

   (c) Hence show that the law of conservation of momentum is valid in the S' frame too if

   $$m_3 + m_4 = m_2 + m_1. \qquad (5.4)$$

   (d) Interpret the last equation.

Your Notes

James Clerk Maxwell
1831-1879

His major achievement was to provide a unified and consistent mathematical formulation of all classical electric and magnetic phenomena using the concept of the 'electromagnetic field'. His equations [38] predicted the existence of electromagnetic waves, and implied light to be an example of such a wave.

6

Waves

If someone strikes a table top at one end, you would be able to feel the vibrations at the other end with your fingertips. Therefore, energy must have propagated from one end of the table to the other without any transfer of matter.

A *wave* refers to the propagation of energy from one point of space to another through adjacent oscillations, without a net transfer of matter. All waves, except electromagnetic and gravitational waves [22], require a material medium for their propagation.

6.1 Types of Waves

Typically[1], waves are characterised as being *transverse* or *longitudinal*. In transverse waves the oscillations are perpendicular to the wave's direction of motion, while in longitudinal waves the oscillations are in the same direction as the propagation of the wave.

An example of a transverse wave is the propagation of energy down a guitar string when one end of it is plucked. Sound waves are examples of longitudinal waves.

6.2 Characteristics

Consider a sinusoidal wave moving along the x-axis, as described by

$$s(x,t) = A \sin 2\pi \left(\frac{x}{\lambda} - \frac{t}{T} \right), \qquad (6.1)$$

---

[1]Surface water waves have a mix of longitudinal and transverse modes.

where $s(x,t)$ is the oscillating quantity characterising the wave (e.g. pressure amplitude for sound or electric field for electromagnetic waves), and t is the time. $s(x,t)$ will be referred to as the 'displacement' since it represents the deviation of the measured quantity from its equilibrium value. The maximum displacement from equilibrium, A, is called the *amplitude* of the wave.

Fig.(6.1) shows the wave at a certain moment in time. Two points along the wave that have the same displacement (including the sign) are said to be in *phase*. The *wavelength* λ is the distance between two adjacent points on the wave that are in phase, for example, the distance between two adjacent peaks.

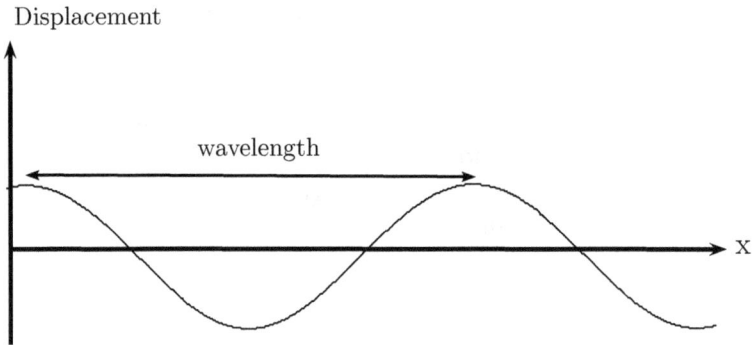

Figure 6.1: Definition of Wavelength.

T is the *period*, the time taken for the wave at a fixed location to return to its previous state. The *frequency*, $f \equiv 1/T$, is the number of oscillations of the wave in one second. So, for a wave propagating at speed v, the wavelength λ, is related to the frequency f, by the relation

$$v = f\lambda. \tag{6.2}$$

When waves propagate from one medium to another, their frequency remains unchanged [23] while their wavelength changes. Thus the speed of waves depends on the medium of propagation.

The energy carried by the wave is proportional to $|A|^2$ where A is the amplitude.

Displacement

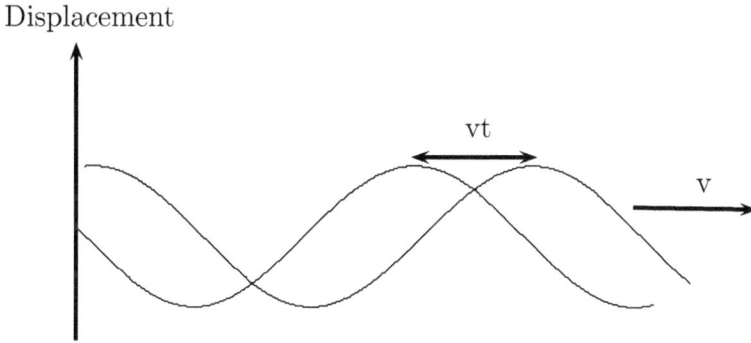

Figure 6.2: Wave moving at speed v.

6.3 Wave Equation

The dynamics of waves is determined by a wave equation. In one dimension, x, the equation describing linear wave propagation is [24]

$$\frac{\partial^2 s(x,t)}{\partial x^2} - \frac{1}{v^2}\frac{\partial^2 s(x,t)}{\partial t^2} = 0, \tag{6.3}$$

with t the time variable and v a positive constant. Here $\frac{\partial}{\partial x}$ is the 'partial derivative' with respect to x keeping t constant, and likewise for the other case and higher order derivatives.

Eq.(6.1) is one particular solution of the wave equation. The general solution is

$$s(x,t) = g(x - vt) + h(x + vt) \tag{6.4}$$

where g and h are any smooth functions of their variables. Values of x and t for which $x - vt = \text{constant}$ correspond to points with the same displacement for the wave g; comparing two such points implies $\Delta x - v\Delta t = 0$, or $v = \Delta x/\Delta t$. Therefore $g(x - vt)$ represents a wave moving to the right with speed v. Similarly, $h(x + vt)$ represents a wave moving to the left with speed v.

Although the general wave (6.4) need not have a simple sinusoidal profile as in Eq.(6.1), a technique called *Fourier analysis* allows us to express a periodic function as the sum of several waves of different periods [25].

6.4   Interference and Superposition

Waves can interact with each other, a process called *interference*. For the *linear* wave equation (6.1), the waves obey the simple *superposition* rule: two waves can combine to form a new wave whose displacement at any (x, t) is the sum of the displacements of the original two waves at the same (x, t).

6.5   Harmonics

If boundaries constrain the propagation of a wave, then its possible wavelengths will be restricted. Consider, as an example, a string fixed at two ends a distance L apart. As the two ends cannot move, we see from Fig.(6.3) that L must be an integer multiple of $\lambda/2$ where λ is the wavelength. So the possible wavelengths are $\lambda = 2L/n$, with n a positive integer.

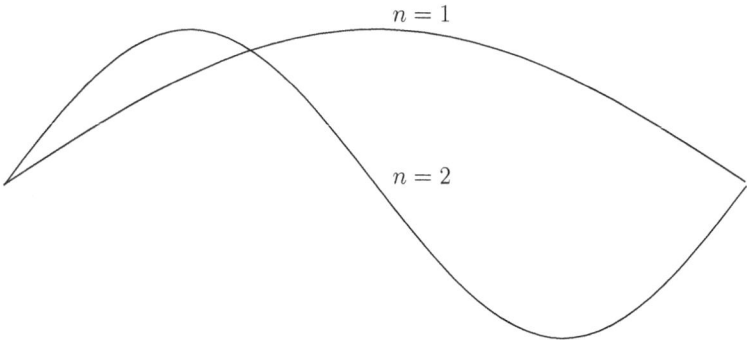

Figure 6.3: Standing waves on a string of length L.

The vibration with $n = 1$ is called the *fundamental mode*, while those with higher n are the *harmonics*.

In this example, the waves are *standing waves*, formed by the interference of waves moving to the left and right: the location of the points of maximum displacement are fixed for each mode.

6.6   Exercises

1. If thunder is heard two seconds after lightning is seen, how far away did the event originate? Device a rule-of-thumb to determine the approximate location of a lightning strike.

2. How do Polaroid sunglasses work? Can one produce 'Polaroid' sound filters?

3. Typically, humans can only hear sound frequencies in the approximate range 20 Hz to 20 kHz. On the other hand, some radio stations transmit at frequencies around 100 MHz. Clarify the apparent discrepancy.

4. How do noise-cancelling headphones work?

5. What is *diffraction*? Give examples from everyday life.

6. As white light passes through a prism, it splits into several colours. Discuss how the speed of the light of different colours (as characterised by their frequency) in the glass is related to their different refractive indices. Calculate also the change in wavelengths as the light passes from air to glass.

7. Why does a wave pulse (for example, a localised disturbance on the water surface) diminish in amplitude the further it propagates from its source?

8. Is it possible to have a wave pulse that keeps its shape as it propagates?

9. If Eq.(6.1) describes a wave as observed from the frame S of Sect.(3.5), find the corresponding equation for observations of the same wave from the frame S' moving with velocity $u = \lambda/T$. Interpret your result.

10. Consider the function $f(x,t) = B\sin(kx)\cos(\omega t)$.

    (a) Show that it is a solution of the wave equation (6.3) if k and ω are related to v.

    (b) Show that $k = 2\pi/\lambda$ and $\omega = 2\pi f$.

    (c) What kind of wave does this represent?

Your Notes

Ludwig Boltzmann
1844-1906

He believed in the reality of atoms long before most other physicists, and developed the field of statistical mechanics to derive the bulk properties of matter from the properties of their atomic constituents. His key achievement was in explaining the Second Law of Thermodynamics as a statistical law, by linking the concept of entropy to the number of microstates corresponding to a given macrostate, Eq.(7.4).

7

Thermodynamics and Statistical Mechanics

In *thermodynamics* one makes statements about the macroscopic physical properties of a system as a whole rather than attempting to describe the states of its many components.

We will mainly be concerned with systems that have reached *thermodynamic equilibrium* (thermal equilibrium), whereby their measurable macroscopic properties, such as volume and temperature (to be defined below), do not change with time.

7.1 Zeroth Law of Thermodynamics

Two systems are said to be in thermodynamic equilibrium with each other if their macroscopic properties remain unchanged when they brought into thermal contact[1].

Zeroth Law: If two systems are each separately in thermal equilibrium with a third system, then they are also in thermal equilibrium with each other.

A practical use of this law is as follows: if some system has a conveniently measurable property, such as the mercury level in a glass tube, it can be designated as a *thermometer* and that property can be used to define *temperature* on a suitable scale. For example, on the Celsius scale, the freezing and boiling points of water at standard atmospheric pressure[2] are labelled as 0°C and 100°C respectively.

---

[1] This means that they are allowed to exchange energy, but not matter, through their boundaries.

[2] Defined as $101,325$ N/m^2.

It follows from the Zeroth Law that systems at the same temperature are in thermodynamic equilibrium.

7.1.1 Gases

A *gas* is a form of matter which occupies the volume of the container it is placed in.

Real gases [26] at moderate pressures and temperatures obey the universal gas law $PV = a(\theta + \theta_0)$, where P is the *pressure*[3] of the gas, V its volume, and θ is its temperature on the Celsius scale. a and θ_0 are universal constants.

An extrapolation of that empirical law to the limit $PV = 0$, gives the value $\theta_0 = 273.15$. It is convenient to define an *absolute temperature scale* T, measured in Kelvin, K, by

$$T = \theta + 273.15 \,. \tag{7.1}$$

The *ideal gas law* is then defined by the equation

$$PV = NkT \,, \tag{7.2}$$

where $k = 1.38 \times 10^{-23}$ J/K is *Boltzmann's constant* and N is the number of molecules in the gas.

Unlike the Celsius, and other ad hoc temperature scales, the absolute temperature has special significance. For example, the absolute temperature T of a gas is proportional to the mean kinetic energy of the molecules making up the gas [27]. The absolute temperature can also be independently defined through the Second Law of Thermodynamics [11].

7.1.2 Radiation

The hotter an object is, the more electromagnetic radiation it emits at shorter wavelengths. Quantitatively, the peak of the energy distribution for an ideal emitter, called a *black body* [28], is located at the wavelength λ_m where

$$\lambda_m T \approx 0.003 \text{ m K} \,, \tag{7.3}$$

which is known as *Wein's displacement law*. So 'blue hot' would be hotter than 'red hot'. The sun has a surface temperature of about 5800 Kelvin, which makes it somewhat 'white hot'.

---

[3]Force per unit area.

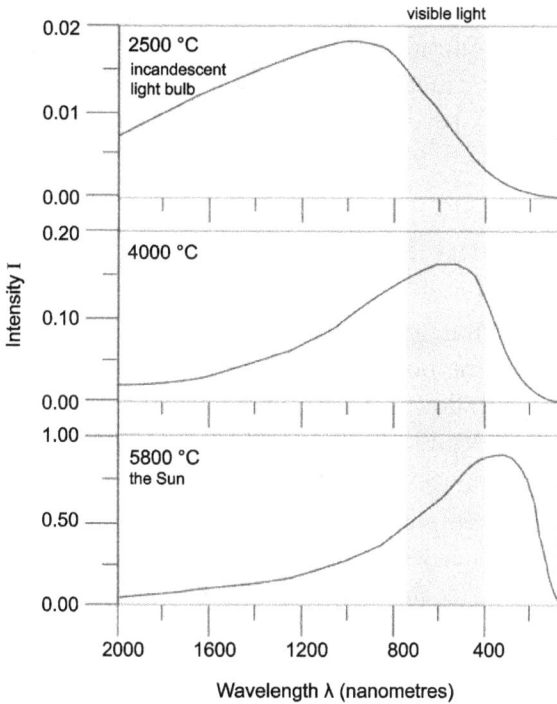

Figure 7.1: Variation of the intensity and peak of radiation from a black body.

7.2 First Law of Thermodynamics

Consider a system consisting of a gas enclosed in a cylinder which has a piston at one end. By pushing the piston in, we do work on the system and thereby transfer energy into it. Alternatively, we could transfer energy into the system by placing the cylinder in contact with another body at a higher temperature. *Heat* is the transfer of energy between two systems due to a temperature difference between them.

First Law of Thermodynamics: The heat input and work done on a system increase its *internal energy*.

The First Law is just our familiar Law of Conservation of Energy, now emphasising the concept of 'internal energy': for a macroscopic piece of matter, it is the energy possessed by the molecules, excluding the bulk energy of the matter. For example, a gas filled cylinder at rest[4] has zero bulk kinetic energy but non-zero internal energy: its centre of mass is at rest, while the molecules are in motion.

7.3   Order and Disorder

Not all processes that are allowed in principle by the First Law of Thermodynamics are possible in practice. For example, we know that heat does not flow spontaneously from a cold body to a hot one. We can understand this using *Statistical Mechanics*, which uses probabilistic techniques to study large systems.

Recalling that temperature is related to the mean kinetic energy of the molecules, then when two bodies are in contact, it is unlikely that the molecules in the hotter body will move even faster while colliding with the slower moving molecules of the colder body. In fact, we naturally expect that due to the molecular collisions, the molecules in the colder body will speed up and those in the hotter body slow down, implying a heat flow from the hot to the cold body.

We also know from experience that it is impossible to construct an engine that would work cyclically by only taking in heat and transforming it completely into mechanical work. Some of the input heat energy must be expelled at a lower temperature[5]. Again, to understand this heuristically, note that internal energy is a disordered form of energy compared to an ordered form such as the bulk kinetic motion of the body. Because the number ($\sim 10^{23}$) of molecules in any macroscopic piece of matter is very large, it is statistically improbable for all the disordered molecules (moving in different directions) to spontaneously act in concert to create maximum bulk kinetic energy.

The impossible processes mentioned here are particular cases of the Second Law of Thermodynamics to be discussed in the next section.

---

[4]The bulk potential energy of the system does not count as part of its internal energy.

[5]For example, a car engine produces hot exhaust gases which are discharged into a lower temperature environment.

7.4 Second Law of Thermodynamics

The *macrostate* of a system is specified by its macroscopic variables such as the volume, pressure, and temperature. On the other hand, specifying the *microstate* of the system means identifying the states[6] of all the molecules composing the system. Clearly many different microstates can correspond to the same macrostate; that is, several different configurations at the micro level can give rise to the same macro variables of volume, pressure and temperature.

In statistical mechanics, the *Boltzmann entropy S* is defined by

$$S = k \ln \Omega \qquad (7.4)$$

where Ω is the number of possible microstates corresponding to the state of the system specified by one macrostate.

Entropy measures our uncertainty about the system, or equivalently the amount of disorder in the system: the larger the number of possible microstates that can correspond to a particular macrostate, the greater our ignorance of the actual microstate, and also the larger the amount of disorder in the system (a system with a limited number of possible microstates is more restricted or ordered).

Second Law of Thermodynamics: The entropy of a closed system never decreases[7].

Consider the following example: an isolated box contains an equal number of two types of molecules, say 'white' and 'black'. The molecules are in constant motion, colliding with each other and with the walls of the box. The volume, temperature and pressure of the gas are constant, defining a fixed macrostate.

Suppose that, at a particular moment in time, we manage to segregate all the white molecules to the right side of the box and the black molecules to the left. Of course, this is an unnatural situation and soon, due to their motion and collisions, the molecules will mix.

One can estimate[8] the entropy of initial and final states as follows. If we specify the location of a molecule simply by whether it is to the

---

[6] Classically, this would mean specifying the positions and momenta of all the particles.

[7] When comparing the final equilibrium state with the initial equilibrium state. Unlike energy, entropy is not conserved.

[8] This is a crude estimate as, among other things, we are ignoring the momenta of the particles in specifying the microstate.

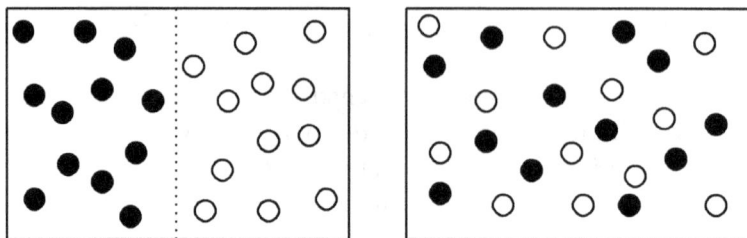

Figure 7.2: Molecules in a box.

left or right of the box, then there is only one initial microstate.

In the final state, since each molecule can be either to the left or right, there are 2^N possible microstates for N molecules. Thus from (7.4) the entropy of the final state is much higher than that of the initial state.

We also know that if the number of molecules is very large, it is extremely unlikely that the gases will revert to the totally segregated state at some future time. Thus the increase in entropy is, for all practical purposes, irreversible.

Note that the irreversibility is not due to the underlying physical laws, but is the result of the system going from an unlikely state to a more probable state; and the fact that for large systems (a large number of molecules) the probability of the system reverting to the ordered state being negligible.

In the above example, the probability of any one molecule being on one side of the box is $1/2$, so the probability that they are segregated is $(1/2)^N$. Even for N as small as 100 this works out to be about 10^{-30}, an infinitesimal quantity. Since macroscopic pieces of matter have N of the order of 10^{23}, the actual probability is much lower.

The Second Law is, therefore, a statement about average behaviour that becomes overwhelmingly likely in a very large system, meaning that exceptions will be unobservable in all practical situations.

Here are other examples: if sugar is placed in a cup of hot tea, it soon dissolves and you will not find a situation whereby the isolated system consisting of the tea and sugar spontaneously separates into its constituent parts. Similarly, if some ink is spilt in a glass of water, it soon spreads and colours the whole glass.

7.4.1   Arrow of Time

In the example above of the gas in a box, the initial configuration is arguably more ordered (organised) compared to the final mixed state. Heuristically then, the Second Law of Thermodynamics implies that the amount of disorder in a closed system tends to increase, and this defines a particular direction in which time 'flows', called the *thermodynamic arrow of time*.

The thermodynamic arrow of time seems to dominate our macroscopic world, see Exercise (10) in Chap.(3).

7.4.2   Information

Boltzmann's formula (7.4) assumes that each microstate is equally probable, a situation that is realisable for an isolated system. However, for a system kept at a fixed temperature (rather than fixed energy), the microstates are not equally likely [29]. In that case one may use the more general *Gibbs entropy* defined by [30]

$$S = -k \sum_i p_i \ln p_i \qquad (7.5)$$

where p_i is the probability for the system to exist in a microstate labelled by i. If each of N possible microstates is equally likely, then $p_i = 1/N$ and Eq.(7.5) reduces to the Boltzmann formula (7.4).

Apart from the Boltzmann constant k, the Gibbs entropy is the same as *Shannon entropy* used in the field of *information theory* [31]. As mentioned earlier, entropy quantifies our uncertainty about the system: the greater the number of possible microstates that can give the same macrostate means that we have less *information* about the actual state of the system at the micro level when presented with only the macroscopic data.

7.4.3   Life

Living systems tend to develop towards greater order, in contrast to the arrow of time dictated by the Second Law of Thermodynamics.

Of course, there is no conflict as the increase in disorder and entropy required by the Second Law refers to closed systems. Living systems are not closed. They use an inflow of energy to drive processes that increase their order (thus decreasing their entropy), and dissipate heat and other waste products that lead to an overall increase in entropy of the universe.

7.5 Third Law of Thermodynamics

Classically, all molecular motion would cease at $T = 0$ (*absolute zero*). However, the *Third Law of Thermodynamics* states that it is impossible to reach absolute zero in practice.

7.6 Spontaneous Symmetry Breaking

Consider Fig.(7.3) which shows a material (the system) in two different states. At the molecular level, the material has 'magnetic domains' which can be thought of as tiny magnets; those are indicated by arrows. Imagine yourself located at the centre of the system.

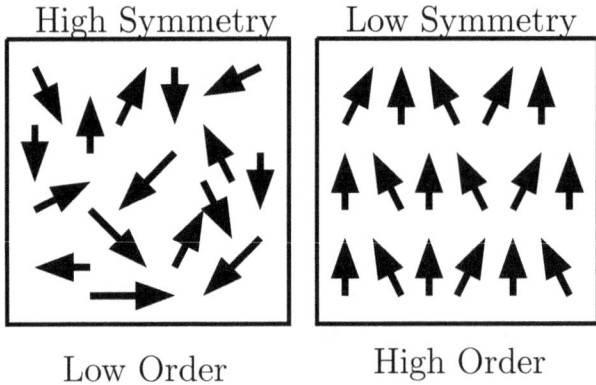

Figure 7.3: A magnetic system in two states. The left state has high symmetry and low order, with zero average magnetism. The right state has low symmetry and high order, displaying macroscopic magnetism.

In the state labelled 'high symmetry', the system would look roughly the same in each direction from your central position: the system has 'rotational symmetry'. On the other hand, the state labelled 'low symmetry' has a preferred direction and will not be rotationally symmetric from your perspective.

However, in terms of the concept of *order* [32], the 'low symmetry' state is clearly more ordered (organised), with the system displaying magnetism at the macroscopic scale. If this ordered state is now heated, the microscopic domains would gain energy, become agitated and disordered. In the process the system would achieve a higher symmetry, but a zero average magnetism.

Conversely, if the material in the non-magnetic state is cooled, the microscopic domains tend to line up: The symmetry is spontaneously broken as the material becomes ordered and magnetic.

The concept of *spontaneous symmetry breaking* is an essential ingredient in the theory of superconductivity, and also in the *Standard Model* of particle physics where it endows fundamental particles with mass.

We see from the example above that 'spontaneous symmetry breaking' may also be phrased as the 'spontaneous generation of order', or *self-organisation*, an idea that is often used in the study of complex systems [2].

7.7 Emergence

An *emergent* property is defined as that which is characteristic of the system as a whole, but not of its component parts [2].

We have seen that the ideal gas law and the Second Law of Thermodynamics are often good approximations when one is talking about the average properties of a system with a large number of molecules. Those laws are examples of *emergent laws*: regularities that are apparent at the level of the system as a whole but are not obvious, or existent, at the molecular level.

Even the concept of temperature is an emergent feature: an individual molecule does not possess a temperature; the absolute temperature T of a gas is proportional to the mean kinetic energy of the molecules making up the gas, and fluctuations from that mean are small only when the number of molecules is very large.

The concept of 'spontaneous symmetry breaking' ('self-organisation') too is an example of 'emergence'. However, the converse need not be true: an emergent property need not be self-organised. For example, the Second Law of Thermodynamics is an emergent law, but it emphasises the creation of disorder rather than order [2].

7.8 Exercises

1. It is sometimes stated that "greenhouse gases trap heat". Explain why radiant energy from the Sun can pass through the atmosphere relatively easily, but radiant energy from the Earth has difficulty escaping.

2. Thermal scanners were often used at airports during the SARS episode to detect people who might have a fever. Explain the physical principle that allows one to deduce temperatures remotely.

3. Estimate the temperature of the glowing charcoal in your barbeque pit.

4. Why is it that you can cool a room using an air-conditioner, but not by using a refrigerator with its door open inside the room?

5. Water frozen in a refrigerator turns to ice in which the motion of the molecules clearly decreases, and thus the amount of disorder (entropy) of the system (water) decreases. Is this a violation of the Second Law of Thermodynamics?

6. The Law of Conservation of Energy states that energy cannot be destroyed but only converted from one form to another. Then what does one mean by statements such as "Do not waste energy" or "Conserve energy"?

7. Solar energy can be used to generate electricity using solar cells or to heat water directly. Which, if any, of those two applications is an inefficient conversion of energy? Explain.

8. The country Xtopia has decided to improve its electrical energy source by building either a hydroelectric power plant or a nuclear power plant. It has the resources to build only one of them. Summarise clearly the advantages and disadvantages of the two options and give your opinion on the best strategy.

9. A web article claimed that scientists had observed a violation of the Second Law of Thermodynamics in a particular system containing a few atoms. Is that possible?

8

Classical Electromagnetism

Electromagnetism is the unified study of the electric and magnetic properties of matter.

8.1 Charges

An atom has negatively charged *electrons* orbiting a positively charged nucleus. A neutral atom has equal amounts of positive and negative charges, but because it is relatively easy to strip the electrons from the outer orbits, one very often ends up with charged atoms. Indeed, this is what happens when one rubs different materials together to form charged objects.

Coulomb's law states that the electrostatic[1] force between two point electric charges Q_1 and Q_2 separated by distance R in vacuum is given by

$$F = \frac{Q_1 Q_2}{4\pi\epsilon_0 R^2} \tag{8.1}$$

where ϵ_0 is a universal constant called the 'permittivity of the vacuum', and the direction of the force is along the line joining the two charges.

An important difference between Coulomb's law and Newton's law of gravitation is that while gravitational force is always attractive, the electrostatic force can be both attractive or repulsive depending on the signs of the charges.

The unit for charge is the *Coulomb* (C). Electric charge comes in integer multiples of the electron charge, which is about 1.6×10^{-19}C [33]. The total charge in a closed system is a conserved quantity.

---

[1] For moving charges there will in general also be a magnetic force, see Eq.(8.3).

8.2 Fields

An electric charge generates an *electric field* in its vicinity which then acts on other charges. The electric field is described by a vector

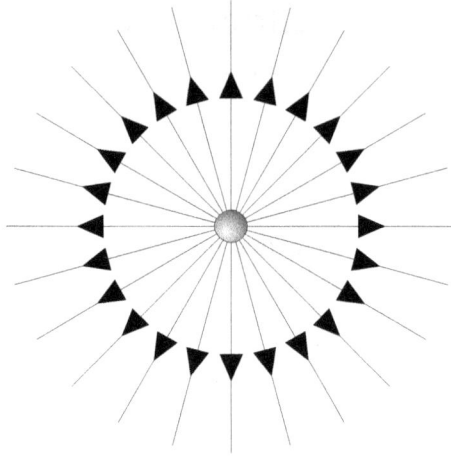

Figure 8.1: Electric field due to a positive point charge.

assigned to every point in space [34].

Mathematically, the electric field produced by a point charge Q is defined as the force it produces on a unit positive *test charge*[2]. Therefore, an electric field $\mathbf{E}$ exerts a force $\mathbf{F}$ on a charge q given by

$$\mathbf{F} = q\mathbf{E} \ . \tag{8.2}$$

The field, therefore, acts as an intermediary in transmitting the influence of one charge on another some distance away. The speed at which information is propagated from one charge to the other is finite, see Sect.(8.6.1). In other words, there is no instantaneous 'action at a distance' that seems to be implied by Eq.(8.1).

8.3 Electric Potential Difference

In an earlier chapter, we encountered the concept of potential energy, an example being gravitational potential energy — it requires work,

---

[2]A test charge is a charge small enough so that its effect on the surroundings is negligible.

done against the force of gravity, to lift an object a certain height, and the object is then said to have gained potential energy.

Similarly, instead of talking of electric forces and fields which are vector quantities, one can introduce the concept of *electric potential difference* between two points, defined as the work required to move one Coulomb of positive test charge between those two points.

Thus the unit of electric potential is Joule/Coulomb and is called the *Volt* (V).

From its definition, we see that in an electric potential field a free positively charged particle will move from a point of higher potential to one of lower potential, while (because of the sign difference) a negatively charged particle will move in the opposite direction.

A *battery* is a device that creates potential differences through chemical means. Chemical potential energy in the battery is used to drive electrons from the positive terminal to the negative terminal, and this continues until the electrostatic potential difference becomes large enough for equilibrium to be reached. At this point, small dry cells have a potential difference of 1.5 V between their two ends.

8.3.1   Current

If a battery is connected by a piece of wire, the potential difference between the two ends causes a flow of electrons. The chemical energy stored in the battery is converted to the energy of the electrons. A *current* of one *Ampere* (A) is the flow of one Coulomb of charge past any point in one second.

Real wires resist the electrical current to varying degrees, due to the collision of the electrons with the atoms of the wire. If I is the current which flows through a wire when the ends are at a potential difference of V, then *Ohm's law* [35] gives the resistance as $R = V/I$. Electrical resistance causes dissipation of energy as heat.

It is important to note that while the electric current travels at the speed of light[3], the net electron flow in one direction is at a much slower *drift velocity* ($\sim$ 0.01 cm/s) because of their collisions with atoms. Also, although there is a current, the wire itself is electrically neutral because of the lattice of positively charged atoms.

A word about conventions. For historical reasons the direction of current drawn in diagrams is that which a positive charge would

---

[3]The speed of electromagnetic waves in the wire is less than in vacuum.

take if it were the carrier: this is called the *conventional current* flow. The actual current due to the motion of electrons is opposite to the conventional current.

8.4 Magnetism

In addition to electric charges and forces, there exist analogous magnetic forces. Magnetic materials always have two opposite poles: a north pole and a south pole, with like poles of materials repelling and unlike poles attracting. Just as we introduced the electric field, we also have the concept of a *magnetic field*, denoted[4] by **B**. The magnetic field lines around a magnet are shown in Fig.(8.2).

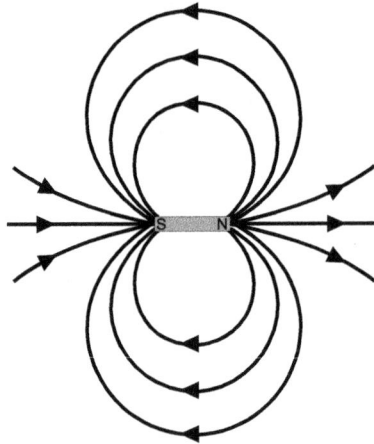

Figure 8.2: Magnetic field lines due to a magnet.

Oersted discovered that electric currents produce magnetic fields. For a straight conductor the direction of the magnetic field is given by the right-hand-rule: if the thumb points along the direction of the conventional current, then the fingers curl in the direction of the magnetic field. The same rule applied to segments of curved conductors can help one to deduce the magnetic field configuration as shown in Fig.(8.4).

---

[4]The magnetic field can be quantified by the Ampere-Maxwell Law [38]. See Fig.(8.3).

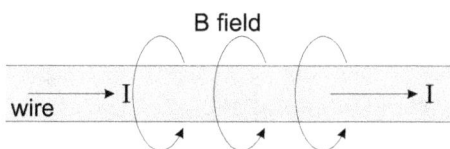

Figure 8.3: Magnetic field lines around a straight conductor. For an infinitely long wire the **B** field forms concentric circles. The magnitude of the field outside the wire is $B = \dfrac{\mu_0 I}{2\pi r}$ where μ_0 is the 'permeability of the vacuum', I the current, and r the radial distance from the centre.

Conversely, Faraday discovered that a time-varying magnetic field produces electric fields. This process is called *electromagnetic induction*. If the magnetic flux[5] through a coil is changed, the induced electric fields produce currents which tend to oppose the change (*Lenz's law*), see caption in Fig.(8.4).

8.5 Lorentz Force

The well-known force between magnets implies, via Oersted's discovery, a force between one magnetic field and a carrier of electric current. Since a current is just moving charge, it is no surprise then that an electric charge moving in a magnetic field experiences a force. A test charge q moving at velocity **v** through a region which has an electric field **E** and a magnetic field **B** experiences an electromagnetic force given by the Lorentz formula [36]

$$\mathbf{F} = q\,(\mathbf{E} + \mathbf{v}\times\mathbf{B})\ . \tag{8.3}$$

This expression makes no reference to 'magnetic charge' [37] as all isolated charges are electrical, and magnetic effects can be understood as due to moving charges.

---

[5]The flux through an infinitesimal area dA with unit normal **n** is given by $d\Phi = \mathbf{B} \cdot \mathbf{n}\, dA$. For a finite area, we integrate over the whole surface.

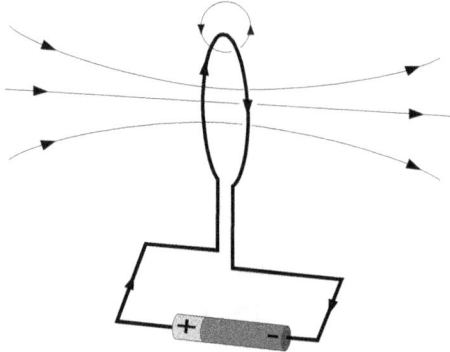

Figure 8.4: Magnetic field lines around a current carrying coil. On the other hand, if the battery is removed but the circuit still closed, then pushing the north pole of a magnet into the coil in the direction indicated by the field lines would generate an induced current which flows in the opposite direction to that shown in the figure so as to oppose the external flux.

8.6 Maxwell's Equations

The various laws of electricity and magnetism are summarised in *Maxwell's equations* [38], which describe the *unified* phenomenon of *electromagnetism*. That is, the electric and magnetic fields are different aspects of the same entity, the *electromagnetic field*, which in general is a function of space and time, and has an associated energy, momentum, and angular momentum.

Electric and magnetic fields appear to be distinct only in certain circumstances. For example, a charge produces only an electric field when observed by someone who is at rest relative to the charge. But from the perspective of a moving observer, it is the charge that is moving, and it will be observed to generate a combination of both electric and magnetic fields (recall that an electric current produces a magnetic field and that current is simply electric charge in motion).

8.6.1 Electromagnetic Waves

Maxwell deduced that disturbances of the electromagnetic field give rise to *electromagnetic (EM) waves*. EM waves are transverse waves, with the oscillating electric and magnetic fields perpendicular to the direction of propagation of the wave.

In contrast to sound waves, no underlying material medium is needed for the propagation of EM waves: that is, EM waves can propagate in a vacuum. Light is just a special case of electromagnetic waves; those with wavelengths in the range of 10^{-6}m to 10^{-7}m form the visible spectrum. All electromagnetic waves travel at the same speed through a vacuum, which is about 3×10^8 m/s.

Frequency (hertz)

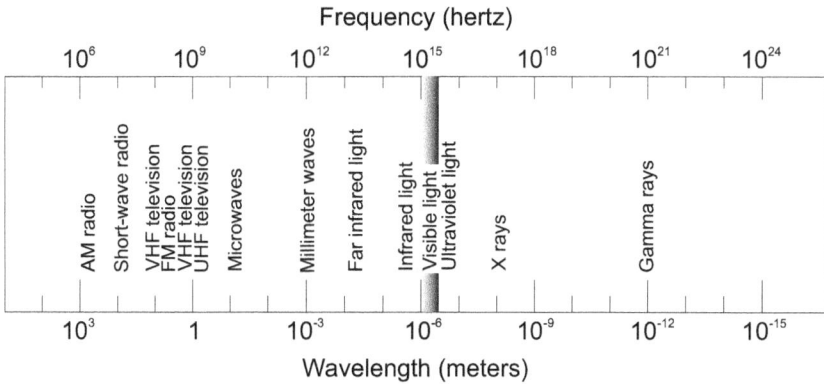

Figure 8.5: The electromagnetic spectrum.

EM waves are said to be *linearly polarised* if the plane in which **E** oscillates is fixed, *circularly polarised* if the plane of polarisation rotates, and *unpolarised* if the **E** field changes direction randomly.

8.6.2   Generation and Propagation of EM Waves

An accelerating electric charge radiates electromagnetic waves. If the electric charge is made to oscillate at a frequency f, as in an antenna, it will produce electromagnetic waves at the same frequency. The waves are sustained by the work done to accelerate the charge.

EM radiation can propagate through empty space because a changing electric field creates a magnetic field, and in turn the changing magnetic field creates another electric field. That mutual feedback sustains the propagation of an electromagnetic wave [38].

Consider for simplicity electric and magnetic fields far from the charges that have generated the radiation. If the electromagnetic wave is propagating in vacuum along the x-direction, then the oscillations of the electric and magnetic field lie in the yz-plane. For example, in a linearly polarised wave, if the electric field $\mathbf{E}$ lies along the y-axis, then the magnetic field $\mathbf{B}$ lies along the z-axis.

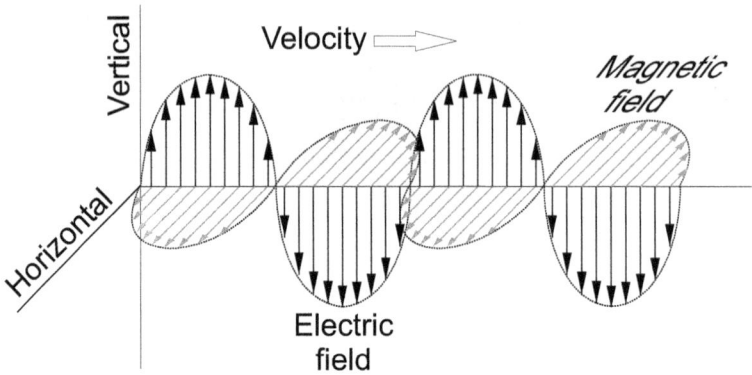

Figure 8.6: Electromagnetic Wave Propagation.

Energy is propagated along the direction of motion of the wave, which in this example is in the x-direction. The *intensity* of the wave is a measure of how much radiant energy per square metre is received at a point each second, and is proportional to $\mathbf{E}^2$. At a large distance R from a small source, the intensity decreases as $1/R^2$, explaining why in many applications transmitters have to be very powerful.

8.7 Exercises

1. Try the following experiment in an environment with low humidity: use a plastic comb to brush your hair for some time and then bring the comb close to (but not touch) some small pieces of paper. Describe what you see and explain the results. Why does this experiment not work if there is considerable humidity?

2. Compare the magnitude of the gravitational force between the electron and proton in a hydrogen atom with the electrostatic force. If electrical forces are so large, why do we not sense them (relative to the gravitational force) in daily life?

3. Some batteries are rechargeable, while most are not. What is the reason for the difference?

4. When electrical power (that is, energy per unit time) P is transmitted at a voltage V along a cable of resistance R, the current that flows is given by $I = P/V$. This current results in some energy dissipation in the form of heat. The power loss P_L is given by $I^2 R$.

   (a) Find an expression for P_L in terms of V, P, R.

   (b) If P is fixed, and R has already been minimised, how can power loss be reduced by adjusting V?

5. What is the difference between 'AC' and 'DC' electrical power? Why is electrical power supplied to homes in the form of AC?

6. Explain why one sees warning signs of "Danger – High Voltage" rather than "Danger – High Current". Why is it that birds can perch on high-voltage lines without getting electrocuted?

7. How does an audio speaker use magnets and electricity to convert electrical signals into sound?

8. Two parallel wires, 10 cm apart, lie in a vacuum chamber. Use Oersted's observation and the Lorentz force equation to determine if the wires would experience a force in each of the situations below. If there is a force, deduce its direction.

   (a) Currents pass through each wire in the same direction.

   (b) The current in one wire is in the opposite direction to that in the other wire.

   (c) Only one wire carries a current.

9. In the S inertial frame, a charge q moving in the horizontal plane at velocity $\mathbf{v}$ passes through a small region which has a magnetic field $\mathbf{B}$ perpendicular to the plane.

   (a) Determine the direction of the force that acts on the charge and its subsequent motion.

(b) Consider now the situation from the frame S' which is attached to the charge. As the charge is at rest in this frame, it cannot experience any magnetic forces. How is this to be reconciled with the conclusion of the last part?

10. Your little brother has hidden your mobile phone inside the microwave oven. The door of the microwave oven is closed, but the oven is not switched on. In an attempt to locate your phone, you call your mobile phone number from another phone, hoping to hear your ringtone. Explain whether that strategy would work (you may assume your mobile phone is switched on).

11. Using common household materials and resources, describe how you can make each of the items below in less than one hour.

(a) A simple compass. (You do not have access to any magnets).

(b) A simple portable electrical power source.

12. An extraterrestrial by the name of Balien has come to seek revenge over his half-brother's demise. The chemical composition of Balien is similar to that of a human. Balien challenges you to a duel. You have the choice of using either a microwave gun, which generates waves around 12 cm, or a UV gun which generates waves around 300 nm. You get to choose first: whichever weapon you pick, Balien is forced to use the other weapon. Also, both of you have two transparent shields to choose from: one made of plastic and the other from glass, but now Balien gets to choose first, and you get what is left. Explain, using clear reasoning, what your choice of weapon should be for survival. (If you lose, the human race will be destroyed. Good luck).

13. How much could your building potentially save in electricity bills if it utilised solar power? How much would it cost to install the solar panels? How long would it take to break even?

14. Show that the wave equation (6.3) is not form invariant under the Galilean transformations Eqs.(3.1-3.2). What are the implications of this non-invariance for electromagnetism?

9

Epilogue

Despite considerable success, the foundations of classical mechanics and classical electromagnetism came under strain by the start of the 20th century. Some of the issues are summarised here.

9.1 Puzzles

1. Maxwell's equations [38] predict that the speed of electromagnetic waves in vacuum is c. In the 19th century it was believed that the speed was c with respect to the 'ether', a special frame at absolute rest; the Galilean velocity transformation (3.5) then implies a different speed for light in other moving inertial frames[1]. But experiments seemed to show that the speed of light was the same in all inertial frames. How could that be?

2. Despite numerous attempts, the formalism of classical physics could not explain why the black-body radiation curve should be of the form shown in Fig.(7.1).

3. At the start of the 20th century, empirical evidence suggested the following model of an atom: a tiny positively charged nucleus with negatively charged particles moving around it. However, an electron in a closed orbit is accelerating since its velocity changes direction even if the speed is constant. Since an accelerating charge always radiates, the electron would continuously lose energy, spiralling into the nucleus in a fraction of a

---

[1]Indeed, the wave equation changes its form under Galilean transformations, see Exercise (14) in Chap.(8).

second. Hence, according to the classical equations of Newton and Maxwell, atoms should be inherently unstable. But you are still reading this sentence. What is wrong?

4. When light is shone on metals, electrons are emitted. According to classical electromagnetism, as the intensity of light is increased, more energy should be transferred from the light waves to the electrons, and so more electrons should be emitted. However, experiments showed that no electrons are emitted if the frequency of the light is below some threshold, regardless of its intensity. Why?

5. Newton's law of gravity (4.13) seems to imply an instantaneous influence of one mass on another, no matter how far that mass is. This seems "magical". Is it possible to reformulate the theory so that information is transmitted at a finite speed through an intermediary field, as in Maxwell's theory?

9.2   Solutions

The first puzzle was solved by Einstein who questioned the foundations of Newtonian mechanics. He replaced the idea of an 'absolute time' by an absolute speed: the speed of EM waves in vacuum was taken to be frame independent. The Galilean transformations were then seen to be approximations of another set, the *Lorentz transformations*[2]. Einstein's Special Theory of Relativity, which has been verified by dozens of experiments, implies, among other things, that time intervals are relative concepts, having different values for observers in relatively moving inertial frames.

It took decades, and the effort of dozens of physicists, to produce a coherent and consistent solution to the second, third, and fourth puzzles [39]. *Quantum Mechanics* describes the behaviour of matter at the microscopic scale using language that is very different from Newtonian mechanics. The deterministic laws of Newton emerge as approximations, valid for describing macroscopic pieces of matter. Likewise, the classical laws of Maxwell are approximations to *quantum electrodynamics*.

---

[2]Correspondingly, the expressions for various quantities, such as momentum, are modified from their Newtonian form.

Quantum theory, which has been verified by countless experiments, implies, among other things, that energy typically appears in discrete units (*quanta*) rather than the continuum inherent in classical mechanics and classical electromagnetism.

Another consequence of quantum theory is that classical statistical mechanics, which deals with distinguishable particles, is an approximation to *quantum statistical mechanics*. The latter takes into account the fact that particles of integral spin, called *bosons*, have dramatically different statistical properties from *fermions*, which are particles with half-integral spin. There is no restriction on the number of bosons that can be placed in a single quantum state accessible to a system, allowing for the possibility of macroscopic manifestation of quantum phenomena as in *superfluidity* and *superconductivity*. On the other hand, at most one (identical) fermion can occupy any single quantum state accessible to a system, and this explains, for example, the shell structure of atoms and their consequent properties.

If the speed of light is the maximum speed of information transfer, as suggested by the Special Theory of Relativity, then a solution of the fourth puzzle was imperative. After a decade of effort, Einstein came up with the General Theory of Relativity, which showed, among other things, that Newton's law of gravitation is an approximation. The General Theory of Relativity has passed all tests to date, and is the main foundation for the field of *cosmology*, the study of the whole Universe.

Joseph-Louis Lagrange
1836-1913

He re-formulated classical mechanics, showing that the equations for the dynamics of a system could be obtained more simply from a variational principle [10]. The 'Lagrangian' approach is also used in the study of classical and quantum fields.

Your Notes

If I have seen further, it is by standing on the shoulders of giants.

Isaac Newton

A

Hints for Selected Exercises

Chapter 2

- Ex.(10b): Make your deduction using the limit $v \to \infty$.

Chapter 3

- Ex.(2c): That is, if $x_1 = x_2$, is it then always true that $x_1' = x_2'$ regardless of the values of t_1 and t_2? See the scenario in Ex.(1).

- Ex.(4): Show that by a rotation and a shift (displacement), the two frames may be aligned.

- Ex.(6): Use Eq.(3.5).

- Ex.(9): In addition to the centripetal component (see exercise in Chap.2), there will be a tangential component if the speed is not constant.

- Ex.(14): See Ref.[8].

Chapter 4

- Ex.(8): Use Newton's law of gravitation and the formula for centripetal acceleration, see Ex.(9) of Chapter 2. For a geostationary orbit, the period of the satellite must match the Earth's rotational period.

- Ex.(10): This implies either that gravity is repulsive at shorter distances or that there is another force that balances gravity. Explore the likely possibilities. See, for example, Chapter 8.

- Ex.(11): From Eq.(4.2), the force required is $F \approx \Delta p / \Delta t$.

- Ex.(16): Use Ex.(2a) from Chap.3.

Chapter 5

- Ex.(2): Potential energy gained in moving up a vertical distance h is $U = mhg$.

Chapter 6

- Ex.(1): The speed of light is much larger than the speed of sound, so the lightning can be taken to indicate the start time of the sound wave from the source.

- Ex.(8): See Ref.[24].

- Ex.(10): See Sect.(6.5).

Chapter 7

- Ex.(1): Estimate the wavelengths of the peak radiations from the Sun and the much cooler Earth using Wein's displacement law. Look up the transmission/reflection properties of those wavelengths in the atmosphere.

- Ex.(6): See Sect.(7.3).

Chapter 8

- Ex.(9b): In the S' frame, the region with magnetic field passes over the charge. A time varying magnetic field creates an electric field (Faraday's law). See also Eq.(8.3).

- Ex.(14): Use $\dfrac{\partial}{\partial x} = \dfrac{\partial}{\partial x'}$ and $\dfrac{\partial}{\partial t} = \dfrac{\partial}{\partial t'} - u\dfrac{\partial}{\partial x'}$. See also Chap.9.

B

Notes and References

[1] *Real World Mathematics*, W.K. Ng and R.R. Parwani (SRI 2014).

[2] *Simplicity in Complexity*, R.R. Parwani (SRI 2015).

[3] *Handbook of Mathematics*, (SRI Books 2016). ebook at www.simplicitysg.net/books/free-ebooks.

[4] In Einstein's *General Theory of Relativity* space and time are part of a four-dimensional *spacetime* whose curved geometry is determined by the energy-mass content. Even when the amount of energy-mass present is negligible, the flat spacetime geometry is *Minkowskian* rather than Euclidean; this is the context for the *Special Theory of Relativity*.
However, Newtonian mechanics, with its absolute Euclidean space and absolute time, is a very good approximation of the macroscopic world when the energy-mass content is low, and the speeds of objects are much less than the speed of light in vacuum.
A simplified account of Einstein's theories, by the master himself, is Ref.[5].

[5] *Relativity: The Special and General Theory*, A. Einstein, at http://www.gutenberg.org/ebooks/30155.

[6] See, for example, *The Nature of Time*, J. Barbour, at https://arxiv.org/pdf/0903.3489.

[7] Exotic forms of macroscopic matter can exist under extreme conditions. For example, neutron stars are composed of neutrons.

[8] In essence, the laws of physics take their 'simplest' form in inertial frames. One may also use Newton's First Law to give an alternative definition of an *inertial frame*, see Sect.(4.1).

[9] The First Law may also be used, in principle, as a consistency check on whether we are using a 'good clock'. First, choose an inertial frame and check that any particle at rest remains at rest, at any point in that frame. That is, identify and exclude forces that could potentially act on the particles. Next, observe if particles can move at constant velocity in that frame. If the velocities change then one may suspect that it is our measurement of velocity that is in error. Assuming that displacement is accurately measured, then there must be a problem with the measurement of time intervals, that is, with the clock (we also assume that we have identified and removed potential velocity dependent forces). We mention this example to point out that though various primary notions might not have an elementary definition, and are sometimes taken in the original formulation to be 'self-evident', the notions do need to form a self-consistent framework.

[10] In advanced treatments of mechanics, a system is more conveniently described by a scalar quantity called the *Lagrangian, L*, and the dynamics is summarised by the *Euler-Lagrange equations* [11]. The Lagrangian approach is compact and more easily generalised to the study of electromagnetic and other fields, both classical and quantum. The connection between symmetries and conservation laws is also more efficiently discussed in the Lagrangian approach.

For a mechanical system with generalised coordinates q_i and their time derivatives $\dot{q}_i$, with $i = 1, 2, \ldots N$, we have $L = T - V$ where T is the kinetic energy and V the potential energy[1]. The Euler-Lagrange equations are

$$\frac{d}{dt}\left(\frac{\partial L}{\partial \dot{q}_i}\right) - \frac{\partial L}{\partial q_i} = 0 \ . \tag{B.1}$$

---

[1]We consider here only those cases where such an identification is possible. There are generalisations to other situations.

These are a set of N coupled second-order differential equations. Now define the generalised momentum p_i by

$$p_i = \frac{\partial L}{\partial \dot{q}_i} . \qquad (B.2)$$

Then

$$\frac{dp_i}{dt} = \frac{\partial L}{\partial q_i} . \qquad (B.3)$$

So, if the Lagrangian is independent of a particular q_j, the corresponding p_j is conserved.

Next, consider the case where the kinetic energy T only depends on the velocities $\dot{q}_i$. Then Eq.(B.3) can be written as

$$\frac{dp_i}{dt} = f_i \qquad (B.4)$$

where we have defined the generalised force by $f_i = -\dfrac{\partial V}{\partial q_i}$. The resemblance of Eq.(B.4) to Newton's Second Law of motion is not coincidental.

Finally, define the *Hamiltonian* of the system by

$$H = \sum_i \dot{q}_i p_i - L , \qquad (B.5)$$

where H is to be expressed as a function of the q_i and the corresponding p_i. For typical cases, $H = T + V$, which is just the total energy of the system. Instead of Eq.(B.1) the dynamics of the system can also be described by *Hamilton's* equations

$$\dot{q}_i = \frac{\partial H}{\partial p_i} ; \qquad (B.6)$$

$$\dot{p}_i = -\frac{\partial H}{\partial q_i} . \qquad (B.7)$$

These are a set of $2N$ coupled first-order differential equations. The Lagrangian and Hamiltonian formulations of dynamics play an important role in the formalism of quantum theory.

[11] *The Feynman Lectures on Physics*, R. Feynman, R. Leighton and M. Sands. Web-book at www.feynmanlectures.caltech.edu.

[12] Fundamental particles such as the electron have an intrinsic angular momentum called *spin*, denoted by **S**. The total angular momentum **J** of a system is then the sum of the orbital and spin parts: $\mathbf{J} = \mathbf{L} + \mathbf{S}$.

[13] In Einstein's General Theory of Relativity [4], gravity is not viewed as a force but rather as the manifestation of a curved spacetime.

[14] Strictly speaking, the masses that appear in Eq.(4.13) are *gravitational masses*. They determine the strength of the mutual gravitational attraction between the particles. Although the gravitational mass has the same units as the inertial mass that appears in Newton's Second Law of motion, the two are conceptually different and in principle need not have the same value. To make this distinction, let us indicate gravitational masses with an asterisk, so that Eq.(4.13) becomes

$$\mathbf{F} = -\frac{GM^*m^*}{r^2}\,\hat{\mathbf{r}}, \tag{B.8}$$

and the acceleration of a particle near the surface of the earth is then

$$g = \frac{m^*}{m}\frac{GM^*}{r^2}. \tag{B.9}$$

Precise experiments have shown that the measured value of g does not depend on the ratio m^*/m, so by an appropriate choice of units we may set $m^* = m$.

The puzzle of why the gravitational mass should be the same as the inertial mass was brilliantly clarified in Einstein's General Theory of Relativity.

[15] More generally, a system is in mechanical equilibrium when both the total external force, and the total external torque, equal zero.

[16] For a smooth force function $F(x)$, a series expansion about $x = 0$ yields $F(x) = c_1 + c_2 x + \ldots$ where the c_i are constants. Equilibrium at $x = 0$ implies $c_1 = 0$. Then, stable equilibrium at $x = 0$ implies that c_2 must be negative (to provide a restoring force).

[17] Eq.(4.15) is an example of a *differential equation* as derivatives of the unknown variable $x(t)$ appear in the equation. The equation

is *linear*, meaning that if x_1 and x_2 are two solutions, then any superposition (sum) of the form $ax_1 + bx_2$ is also a solution for any constants a and b. The linear equation is often an approximation to a nonlinear equation which does not have the simple superposition property. For example, the equation for the large angle oscillations of a simple pendulum is nonlinear [19].

[18] In realistic cases one must include resistive (*damping*) forces which dissipate the energy. For weak damping, the motion would still be approximately periodic, but the amplitude of oscillations would decrease unless an external driving force maintains it.

[19] The period for large angle oscillations of a simple pendulum depend on the amplitude. See, for example, *An Approximate Expression for the Large Angle Period of a Simple Pendulum*, R. Parwani, at https://arxiv.org/abs/physics/0303036.

[20] In Special Relativity, the energy of a particle moving at velocity v is $E = \gamma mc^2$, where m is the mass of the particle, c the speed of light in vacuum, and $\gamma = 1/\sqrt{1 - v^2/c^2}$. For a particle at rest, $E = mc^2$, so mass is actually a form of (potential) energy. As the value of c is large, even tiny amounts of mass correspond to large amounts of energy.

For a system of particles, the system mass M is defined through $(Mc^2)^2 = E^2 - (\mathbf{P}c)^2$ where E and $\mathbf{P}$ are the total energy and momentum of the system. As E and $\mathbf{P}$ are conserved, so M is conserved too, but it is not simply the sum of the masses of the component parts.

For example, consider a mass M at rest and choose for simplicity units where $c = 1$. The energy of the system is then the same as the system mass $E = M$. If this mass explodes (as in nuclear *fission*), breaking into two pieces of mass m_1 and m_2 which move in opposite directions, then conservation of energy gives $E = E_1 + E_2$, which translates to $M = \gamma_1 m_1 + \gamma_2 m_2$. This last expression can be re-written as $M - (m_1 + m_2) = (\gamma_1 - 1)m_1 + (\gamma_2 - 1)m_2$, and since the right-hand side is greater than zero (because the 'gamma' factors are larger than 1), we see that M is not simply the sum of m_1 and m_2. The difference manifests as other forms of energy, such as kinetic energy: it is in this sense that one says "mass can be converted to energy", and vice versa.

[21] Maxwell's equations [38] can be more conveniently re-written in terms of a scalar potential ϕ and a vector potential $\mathbf{A}$. Then, for example, $\mathbf{B} = \nabla \times \mathbf{A}$ and Eq.(B.16) is automatically satisfied. Indeed, using the potential functions seems to be the only way to get a simple Lagrangian for electromagnetism and to make the transition to quantum theory with matter fields ψ. In terms of the potential functions, the equations have a symmetry called *gauge invariance*, which is the symmetry behind charge conservation.

[22] An electromagnetic wave propagates by mutually reinforcing electric and magnetic fields, see Sect.(8.6.1). A *gravitational wave* is a ripple in the curvature of spacetime, as predicted by Einstein's General Theory of Relativity [4, 13].

[23] Consider the interface between two media through which a wave is propagating from left to right. The vibrations of the medium slightly to the left of the interface must, by continuity, match the vibrations immediately to the right of the interface. Thus the frequency of the wave remains the same as it transits. Its speed depends on the inertial ('resistance to motion') and elastic ('stretchability') properties of the medium. For example, the speed of transverse waves on a vibrating string is $\sqrt{T/\rho}$ where T is the tension of the string and ρ its linear mass density.

[24] This is an example of a *partial differential equation* with two independent variables x and t. The version in three-space dimensions would have four independent variables. This equation is linear [17], and so satisfies the superposition property. Linear wave equations are often approximations to nonlinear wave equations: the later may have novel solutions called *solitary waves*, which are wavepackets (pulses) that maintain their shape as they propagate, unlike wavepackets of linear equations which tend to disperse with time.

[25] For a continuous periodic function $s(z) = s(z + L)$, we have its *Fourier Series* representation

$$s(z) = \sum_{n=0}^{\infty} A_n \sin\left(\frac{2\pi n z}{L} + \phi_n\right) , \qquad (B.10)$$

where the coefficients A_n and ϕ_n depend on certain integrals over $f(z)$.

[26] The *Van der Waals* equation gives a better description than the ideal gas law for gases at low temperatures or high pressures. It takes into account the finite size of molecules and their interaction with each other,

$$\left(p + \frac{a}{v^2}\right)(v - b) = kT ,$$
(B.11)

where v is the volume divided by N (the number of molecules) and a, b are constants.

[27] In *Kinetic Theory*, macroscopic properties of matter are derived from detailed models of molecules and their interactions.

[28] Perfect black-body radiation can be observed coming out of a tiny hole of a cavity kept at a temperature T; the spectrum of the radiation depends only on T. The continuous spectrum emitted by other large bodies is often approximately treated as black-body radiation.

[29] Specifying the possible microstates of an isolated system when its energy is known and fixed gives rise to the *microcanonical ensemble*. On the other hand, if only the temperature of the system is fixed, through contact with a large heat reservoir, one arrives at the *canonical ensemble*. In a canonical ensemble the system can exchange energy with the reservoir, but the average energy of the system is kept fixed, see [31].

[30] The Gibbs entropy is equivalent to entropy as defined at the thermodynamics level. Consider a process whereby a very small quantity of heat ΔQ is added reversibly[2] to a system at fixed temperature T. Then the entropy change of the system is defined by

$$\Delta S = \frac{\Delta Q}{T} .$$
(B.12)

As a consistency check of the above expression, consider the example of an isolated system consisting of two bodies in contact:

---

[2]That is, the process is reversible in principle, whereby the system retraces its path back to its original state.

a large hot body at temperature T_2 in contact with a colder large body at temperature T_1. If the hot body loses a small amount of heat ΔQ, then its entropy change is

$$\Delta S_2 = \frac{-\Delta Q}{T_2},$$

while the entropy of the colder body, which absorbs that same amount of heat, increases by

$$\Delta S_1 = \frac{\Delta Q}{T_1}.$$

The entropy change of the whole system is therefore

$$\begin{aligned}
\Delta S &= \Delta Q \left(\frac{1}{T_1} - \frac{1}{T_2} \right) \qquad \text{(B.13)} \\
&= \Delta Q \left(\frac{T_2 - T_1}{T_1 T_2} \right) \\
&> 0.
\end{aligned}$$

The net entropy of the system has increased because of the heat flow from the hot body to the colder one. Thus using the expression (B.12) gives a result that is consistent with common experience (the direction of heat flow), and the Second Law.

[31] One application of the Gibbs-Shannon entropy is in Jaynes' *Principle of Maximum Entropy (Uncertainty)*, which states that the theoretical probabilities characterising the microstates of a system (not necessarily in thermodynamics) should be chosen in such a way as to maximise the entropy while satisfying given constraints. For example, if the mean energy is fixed, $\bar{E} = \sum_i E_i p_i$, with the index i labelling the microstates, then maximising the Gibbs-Shannon entropy under variations of the probabilities p_i yields the canonical distribution $p_i \propto \exp(-\beta E_i)$, where β is the Lagrange multiplier enforcing the mean energy constraint.

[32] In the study of *phase transitions*, such as when a substance changes from the liquid phase to a gaseous phase, the 'order' in the system is characterised by an *order parameter* whose value changes significantly across the phase transition. For a magnetic system, the net magnetisation is the order parameter.

[33] *Quarks*, which are the constituents of protons and other particles called 'hadrons', carry fractional electric charge (in units of
the charge of the electron). However, quarks are tightly bound
inside hadrons by the strong force, and no isolated quarks have
been observed. The hadrons themselves carry an integral electric
charge.

[34] Not all fields are vector fields. For example, assigning a temperature to each point of space leads to a scalar field, while in the
General Theory of Relativity one uses 'tensor fields'.

[35] The resistance of a piece of wire depends on the temperature.
Ohm's law is an empirical relation (emergent law) that is approximately obeyed by many, but not all materials. Under certain conditions even common materials will display deviations
from Ohm's law.

[36] In the Lorentz force formula the fields are those produced by
external entities, and do not include those generated by the test
charge. Furthermore, if the charge q is accelerating, it will emit
electromagnetic radiation, and its energy loss can be effectively
described as due to a 'radiation reaction force' [11].

[37] Isolated magnetic poles, called *magnetic monopoles*, have not
been observed. The non-occurrence of magnetic monopoles in
Nature is somewhat of a mystery for modern theoretical physics
as almost all of the unified theories of fundamental forces predict
their existence. Thus understanding why they are not seen might
require some new ideas.

[38] In vector notation, Maxwell's equations are a set of four par-
tial differential equations[3] for the fields $\mathbf{E}(\mathbf{r}, t)$ and $\mathbf{B}(\mathbf{r}, t)$. They
involve the vector differential operator ∇ which in Cartesian co-
ordinates with the usual unit vectors takes the form

$$\nabla = \mathbf{i}\frac{\partial}{\partial x} + \mathbf{j}\frac{\partial}{\partial y} + \mathbf{k}\frac{\partial}{\partial z} \ . \tag{B.14}$$

---

[3]They may also be written in integral form.

The equations are

$$\nabla \cdot \mathbf{E} \;=\; \frac{\rho}{\varepsilon_0} \; ; \tag{B.15}$$

$$\nabla \cdot \mathbf{B} \;=\; 0 \; ; \tag{B.16}$$

$$\nabla \times \mathbf{E} \;=\; -\frac{\partial \mathbf{B}}{\partial t} \; ; \tag{B.17}$$

$$\nabla \times \mathbf{B} \;=\; \mu_0 \left(\mathbf{j} + \varepsilon_0 \frac{\partial \mathbf{E}}{\partial t} \right) . \tag{B.18}$$

The first equation is *Gauss's Law* with ρ the electric charge density, and ε_0 is a constant called the 'permittivity of the vacuum'. The second equation implies the absence of magnetic monopoles. The third equation is *Faraday's Law of Induction*. The fourth equation is the *Ampere-Maxwell Law* with $\mathbf{j}$ the electric current density, and μ_0 is a constant called the 'permeability of the vacuum'.

The time-variation of the electric field in the last equation was added by Maxwell to the original expression by Ampere so as to ensure *electric charge conservation*, which can be summarised by the equation

$$\frac{\partial \rho}{\partial t} + \nabla \cdot \mathbf{j} = 0 . \tag{B.19}$$

In a vacuum, where $\rho = \mathbf{j} = 0$, Maxwell's equations lead to the *wave equation*

$$\nabla^2 \mathbf{V} - \frac{1}{c^2}\frac{\partial^2 \mathbf{V}}{\partial t^2} = 0 \tag{B.20}$$

where $\mathbf{V}$ stands for either $\mathbf{E}$ or $\mathbf{B}$, and $c = 1/\sqrt{\varepsilon_0 \mu_0}$ is the speed of the electromagnetic waves in vacuum.

The electromagnetic waves predicted by Maxwell were later observed in a laboratory by Hertz. This example, whereby Maxwell modified Ampere's law to make it consistent with a desire for charge conservation, and ended up predicting electromagnetic waves which were later detected, is a classic example of the *scientific method*. As a bonus, light was understood to be an electromagnetic wave.

[39] *Thirty Years That Shook Physics*, G. Gamow (Dover 1985).

Index

Physics
Volume One: Classical Foundations
ISBN: $978-981-11-3362-6$ (pbook)
ISBN: $978-981-11-3363-3$ (ebook)
Companion website: www.simplicitysg.net/books/physics

The Author
Dr. Rajesh R Parwani is a theoretical physicist, with interests in quantum theory, cosmology, astrobiology, consciousness, and yoga. He currently resides on the Third Rock from the Sun.

Webpage: www.simplicitysg.net
Facebook: www.facebook.com/srisg/
Email: enquiry@simplicitysg.net

www.ingramcontent.com/pod-product-compliance
Lightning Source LLC
Chambersburg PA
CBHW060633210326
41520CB00010B/1589